Ecology and Management of Appalachian Ruffed Grouse

Dedicated to the memory of Tom Allen,

naturalist extraordinaire.

Ecology and Management of Appalachian Ruffed Grouse

Dean F. Stauffer,
Editor

John W. Edwards, William M. Giuliano and Gary W. Norman,
Associate Editors

That all who know the grouse come to have an affection for this unpredictable thunderbolt of the uplands, is axiomatic. Its rolling drum and roaring flush add a personality to the woods which nothing can replace.

—Gardiner Bump, Robert W. Darrow, Frank C. Edminster, and Walter F. Crissey, *The Ruffed Grouse,* 1942

ISBN 978-0-88839-667-9

Cataloging in Publication Data

Ecology of Appalachian ruffed grouse /
Dean F. Stauffer, editor ; John W. Edwards, William Giuliano
and Gary W. Norman, associate editors.

Includes bibliographical references.
ISBN 978-0-88839-667-9

1. Ruffed grouse—Appalachian Region. 2. Ruffed grouse—Ecology—Appalachian Region. I. Stauffer, Dean F. (Dean Fiske), 1953–

QL696.G27E36 2010 598.6'350974 C2010-905978-6

Printed in South Korea — PACOM

Copy Editor: Theresa Laviolette
Production: Ingrid Luters
Cover Design: Ingrid Luters

Front cover photograph by Harold L. Jerrell
Back cover image: Robert Long

Published simultaneously in Canada and the United States by

HANCOCK HOUSE PUBLISHERS LTD.
19313 Zero Avenue, Surrey, B.C. Canada V3S 9R9
(604) 538-1114 Fax (604) 538-2262

HANCOCK HOUSE PUBLISHERS
1431 Harrison Avenue, Blaine, WA U.S.A 98230-5005
(604) 538-1114 Fax (604) 538-2262

Website: **www.hancockhouse.com**
Email: **sales@hancockhouse.com**

CONTENTS

Contributors

Mark A. Banker

Mark has been the regional wildlife biologist for the Ruffed Grouse Society in the southern Appalachian Mountains region since 1998. He earned his BS and MS in wildlife science from Penn State and Virginia Tech. His main areas of interest are conservation education and implementation of forest habitat management.

David A. Buehler

Dave is a professor of wildlife science in the Department of Forestry, Wildlife and Fisheries at the University of Tennessee–Knoxville. He received his Bachelor of Science degree and his Master of Science degree in wildlife ecology from the University of Wisconsin–Madison and his doctorate in wildlife science from Virginia Polytechnic Institute and State University. Dr. Buehler's research interests are focused on avian-habitat relationships, including developing new management strategies for species in need of management.

George B. Bumann

George was fortunate enough to work as a master's degree candidate at Virginia Tech in association with the ACGRP. He is now a sculptor and naturalist working in Yellowstone National Park where he lives with is wife Jenny and two black Labradors. He teaches ecology and art courses for the Yellowstone Institute, as well as for other organizations in that region and across the nation. His bronze sculptures can be found in collections throughout the United States and England, and in the National Museum of Wildlife Art in Jackson Hole, WY.

Daniel R. Dessecker

Dan is the director of conservation policy for the Ruffed Grouse Society. As such, he is responsible for the administration and development of the society's conservation policy and land management initiatives in North America. He received a BS (1981) from the University of Wisconsin–Stevens Point and an MS (1984) from the Pennsylvania State University. His professional interests include the historical and current roles of disturbance in forest ecology, and he has published extensively on this and related topics.

Patrick K. Devers

Pat works for the U.S. Fish and Wildlife Service as the science coordinator for the Black Duck Joint Venture. He earned his PhD from Virginia Tech studying the population ecology of ruffed grouse in the southern and central Appalachians. His interests include wildlife population ecology and the human dimensions of natural resources conservation.

Carrie (Schumacher) Dobey

Carrie Schumacher Dobey graduated from the University of Tennessee in 2002 with a Master of Wildlife Science. Her thesis was entitled "Ruffed Grouse Habitat Use in Western North Carolina." While completing her master's, she worked as a wildlife technician on the elk reintroduction into the Great Smokey Mountains National Park. She then worked as a wildlife biologist at Ft. Campbell, KY, and later moved to Wyoming where she now serves as a regional terrestrial habitat biologist.

Chris A. Dobony
Chris is currently a fish and wildlife biologist on Fort Drum Military Installation in Fort Drum, New York. He assists in the management of all natural resources on the installation, with specific focus on baseline species surveys and deer and beaver management. He received his BS in environmental forest biology from the SUNY College of Environmental Science and Forestry (1997), and his MS in wildlife and fisheries resources from West Virginia University (2000).

John W. Edwards
John is an associate professor in the Wildlife and Fisheries Resources Program, Division of Forestry and Natural Resources at West Virginia University. He earned a BS in wildlife management at WVU, and received his MS and PhD degrees at Clemson University. His professional interests focus on forest wildlife habitat relations, management of threatened and endangered species, and the preservation of our hunting heritage.

Erik R. Endrulat
Erik originated from southwestern Connecticut. He attended the University of Rhode Island for his BS and MS degrees in environmental science and wildlife biology. He has worked for the North Carolina Wildlife Resources Commission, the RI Division of Fish and Wildlife, Harvard Museum of Comparative Zoology. His current position is with the Rhode Island Natural History Survey where he does GIS analysis and web development.

Todd M. Fearer
Todd received his BS in wildlife and fisheries science from Penn State in 1995 with minors in forest science and international agriculture, and completed his MS and PhD in wildlife science at Virginia Tech. His research interests focus on the conservation of oak forest ecosystems, wildlife-habitat relationships at the landscape scale, landscape ecology, and geographic information systems.

William M. Giuliano
Bill is a professor and wildlife extension specialist in the Department of Wildlife Ecology and Conservation, University of Florida. He earned a PhD in wildlife science in 1995 from Texas Tech University, where he studied the reproductive ecology of northern bobwhite and scaled quail. His teaching, research, and extension interests include integrating wildlife management in to other land uses.

Craig A. Harper
Craig is an associate professor and the extension wildlife specialist in the Department of Forestry, Wildlife and Fisheries at The University of Tennessee. He received his PhD from Clemson University where he worked with wild turkeys in the mountains of North Carolina. Craig is responsible for developing wildlife-related programs for UT Extension and assisting extension agents with matters concerning wildlife throughout the state. Craig remains active in research with on-going programs in quality deer management, forest management for ruffed grouse and wild turkeys, and applied habitat management for various wildlife species, including timber stand improvement, native warm-season grasses, and food plot establishment and management.

Scott Haulton
Scott was born and raised in upstate New York, where he received his undergraduate training from State University of New York College of Environmental Science and Forestry. After completing graduate studies at Virginia Tech in 1999, Scott worked with forest preserve agencies in Illinois and the Adirondack Ecological Center in New York, studying and managing wildlife populations in managed forests. Presently, his professional interests include suburban deer management and the restoration of native communities in Midwest natural areas.

Benjamin C. Jones
Ben holds a PhD in natural resources from the University of Tennessee. Ben also received an MS in wildlife and fish-

eries science from Mississippi State University, and a BS in wildlife and fisheries science, with a forest science minor, from Pennsylvania State University. Professional interests include impacts of silvicultural prescriptions on wildlife and the use of forest management to improve wildlife habitat. Currently, Ben is employed by the Pennsylvania Game Commission where he is responsible for developing habitat management plans for wildlife on public lands.

Roy L. Kirkpatrick

Roy is a professor emeritus of wildlife science at Virginia Tech. He dedicated his professional career to the nutritional ecology of mammals and birds, including many studies on ruffed grouse. He now dedicates his retirement career to his wife, children, and grandkids.

Jennifer (Fettinger) Kleitch

Jennifer received her bachelor's degree in wildlife management in 1999 from Michigan State University and her master's degree in wildlife science with a minor in statistics from the University of Tennessee in 2002. After obtaining her graduate degree, she conducted research with the Michigan Natural Features Inventory and later gained employment with the Michigan Department of Natural Resources Wildlife Division in Lansing, Michigan.

Robert Long

Bob worked with the ACGRP from 1999–2002 conducting fieldwork on the Pennsylvania study site and researching ruffed grouse nutrition and condition. He holds a BS in wildlife science from Virginia Tech and an MS from West Virginia University. He is currently the wild turkey and upland game bird project manager for the Maryland Department of Natural Resources.

Gary W. Norman

Gary was the forest game bird project leader for the Virginia Department of Game and Inland Fisheries (VDGIF). Gary received a BS at West Virginia University and MS at Virginia Tech. Previously he worked as a biologist with the West Virginia Division of Natural Resources. His primary professional interests are population dynamics and habitat management of ruffed grouse and wild turkeys. He served as coordinator of the Appalachian Cooperative Grouse Research Project.

Joy O'Keefe

Joy is a doctoral student at Clemson University, with a BS from NC State and MS from Eastern Kentucky University. As part of the ACGRP, Joy enjoyed trapping and tracking grouse in Kentucky and West Virginia in 2000–2001, and researching use of elevation by grouse for her thesis. Now Joy is studying the relationship between forest management and bat ecology and is active in another cooperative conservation organization, the Southeastern Bat Diversity Network.

Aaron Proctor

Aaron served as a technician at the ACGRP Virginia-2 study site in the summers of 2001 and 2002. He graduated from Virginia Tech in 2003 with a BS in wildlife science, and finished an MS in wildlife and fisheries resources at West Virginia University in 2010, where he studied effects of nutritional deficiency and body condition on reproduction in female ruffed grouse.

Brian W. Smith

Brian is the Appalachian Mountains Joint Venture Coordinator, located in Blacksburg, Virginia. He earned his PhD from West Virginia University, studying nesting behavior, brood ecology, and juvenile dispersal of ruffed grouse as part of the Appalachian Cooperative Grouse Research Project. His interests focus on habitat management and restoration for rare and declining species, and policy issues with regards to avian conservation.

Dean F. Stauffer

Dean is a professor of wildlife sciences and Associate Dean for Academic Programs in the College of Natural Resources and Environment at Virginia Tech, where he has been since 1983. He earned his PhD from the University of Idaho, studying habitat ecology of blue and ruffed grouse in southeastern Idaho. His interests center on habitat relationships of wildlife, and evaluating and managing habitat.

David E. Steffen

Dave has been the forest wildlife program manager for the Virginia Department of Game & Inland Fisheries since 1988. Previously he was a biologist with the Mississippi Department of Fisheries, Wildlife, & Parks. Dave earned BS and MS degrees in fisheries and wildlife sciences from Virginia Tech and a Master of Applied Statistics from Louisiana State University. He is a certified wildlife biologist and past-president of the Virginia chapter of The Wildlife Society. His interests include population dynamics and human dimensions with an emphasis on game management applications.

David A. Swanson

Dave is forest wildlife research supervisor for the Ohio Division of Wildlife, where he has been since 1993. He earned his PhD from West Virginia University, studying the population dynamics of wild turkey hens. His interests include evaluating the effects of habitat management on forest wildlife and population dynamics.

Brian C. Tefft

Brian is a wildlife biologist with the Rhode Island Department of Environmental Management–Division of Fish and Wildlife working at the Great Swamp Field Headquarters, Rhode Island. Brian earned a BS in natural resources management from the University of Rhode Island in 1978 and an MS in wildlife management from University of Maryland–Frostburg State University, in 1981. He is involved with research and management on wild turkeys, ruffed grouse, and rabbits, and has managed and studied wetland and salt marsh restoration projects, and the enhancement of these wildlife habitats in Rhode Island. His research interests also include management of early successional habitats, and effects of forest management and age class diversity on wildlife at the landscape scale.

John M. Tirpak

John received a BS in wildlife resource management from West Virginia University in 1996. Following graduation, John held a number of temporary field positions in West Virginia, Ohio, Louisiana, and Connecticut, mostly working on habitat associations of breeding landbirds. In 1999, John began pursuing an MS at California University of Pennsylvania. There he began studies on the influence of habitat on ruffed grouse reproductive ecology as part of the Appalachian Cooperative Grouse Research Project. In 2001, John entered the PhD program at Fordham University, where he continued his research on ruffed grouse and built spatially explicit population models of ruffed grouse on seven sites throughout the Appalachian region. Currently, John is a post-doctoral fellow at the University of Missouri–Columbia and is stationed at the Patuxent Wildlife Research Center/Lower Mississippi valley joint venture office in Vicksburg, Mississippi.

Darroch M. Whitaker

Darroch completed a BS in resource conservation (McGill University, 1994), an MS in animal behavior (Memorial University, 1997), and a PhD in wildlife science (Virginia Tech, 2003). His research focuses on the habitat ecology, spatial ecology, and demographics of forest birds. He currently is an ecologist with Parks Canada, working in Gros Morne National Park in Newfoundland.

Preface

The ruffed grouse is a popular game bird wherever it occurs. This forest grouse is ardently pursued by hunters throughout its range, and there are those who would argue it is indeed the "king of gamebirds." The quickening of our heart at the rise of a grouse having been pointed by your favorite dog is an experience that is not soon to be forgotten. Likewise, the first ethereal sound of a drumming male floating though the woods in the spring has the ability to stir our blood and affirms that spring truly is not far off.

Throughout their range, ruffed grouse show an affinity for early successional, young forests, and are particularly associated with the distribution of aspen across the U.S. and Canada. Aspen stands provide highly nutritious food for grouse during winter, and habitat conditions that allow them to thrive throughout the year. Most of the research that has been conducted on ruffed grouse ecology and management has been from the core of their geographic range where aspen is common, and grouse populations are high. However, life is different for ruffed grouse on the periphery of their geographic range, where aspen is uncommon or absent.

In the northern portions of their range, ruffed grouse are subject to a 10-year cycle in population abundance; in the southern latitudes where populations generally exist at lower densities, we do not see detectable 10-year population cycles. Additionally, as we move to the fringes of their distribution, they become more dependent on herbaceous vegetation and buds of trees and shrubs other than aspen. Relatively little work has been done on the ecology of ruffed grouse outside their core range. In the central and southern Appalachian Mountains, there have been a few studies conducted, but not nearly to the extent of the work done in the Lake States area of their range. Results from the U.S. Fish and Wildlife Service's Breeding Bird Survey indicate that grouse populations from central Pennsylvania southward have declined over the last three decades. State wildlife agencies were concerned about the decline in grouse populations and grouse hunters. Grouse hunters and outdoor writers began to question the impacts of hunting, particularly late season, on grouse populations. Some states responded to this concern by shortening their grouse hunting seasons. Virginia, for example, shortened their grouse season from February 28 to February 15 in 1972. The season was again shortened, to January 31, in 1977. Uncertainty remained about the impacts of these season reductions in Virginia as the Department of Game and Inland Fisheries Board lengthened the season to the second Saturday in February in 1989.

Thus, by the mid 1990s, the time was ripe for an intensive study of ruffed grouse that would enhance our understanding of their ecology and provide guidance for their management in the Appalachian Mountains. In spring of 1996, the Appalachian Cooperative Grouse Research Project (ACGRP) was born. The ACGRP brought together wildlife researchers and managers from Kentucky, Maryland, North Carolina, Ohio, Pennsylvania, Rhode Island, Tennessee, Virginia, and West Virginia who were interested in the status of grouse, and willing to cooperatively establish a study to investigate more thoroughly their ecology and management in this region. Data collection for this project began in 1996 and concluded in fall 2002. Research was conducted on 12 study sites distributed across eight states, and involved the cooperative efforts and support of scores of individuals from a diversity of state agencies, federal agencies, universities, and private organizations.

Much of the information generated from this project is buried in graduate student theses and dissertations, and annual reports to various agencies. Additionally, we have jointly presented dozens of talks at professional meetings on the results, and published numerous papers in scientific journals. The results that describe the ecology of this intriguing and fascinating bird thus have been reduced to dry, scientific prose with arcane statistical descriptions of just how these birds survive and relate to their environment. Although the hard-core scientific presentations of our results are necessary and help to establish the scientific credibility of this work, make the graduate students more employable, and help to ensure the promotion, tenure, and pay raises of faculty, they strip away some of the personality of the bird.

There is an interesting story to be told concerning the ecology and management of ruffed grouse in the Appalachians. Our goal is to tell that tale in a way that is informative and educational to everyone. We expect that all who enjoy ruffed grouse, whether hunters, naturalists, birders, or researchers may benefit from our efforts here. The myriad theses, dissertations, presentations, and publications that have come from the ACGRP tell only a part of the story. Here we strive to pull together all that we have learned about grouse into one place. For clarity, we have stripped away most of the gory, statistical treatment of the data, but we have in no way tried to minimize our results. The results presented here draw heavily from the theses and dissertations of the graduate students who conducted the majority of the research; these works are listed in the appendix. While the geographic extent of our study sites encompassed the central and southern Appalachians, throughout this book we will refer simply to the "Appalachians" or "Appalachian grouse" when referring to the area or the bird.

This book comprises several sections that summarize major components and contributions of the ACGRP. Authors of each chapter represent the ACGRP members who were most involved in that particular component of the work. The first two chapters set the stage for the larger project and results. In Chapter 1 you will find an overview and introduction to the ACGRP and Chapter 2 describes the general methods used throughout the study. The next four chapters pertain to nesting, reproduction, and survival of ruffed grouse. In Chapter 3 you will find a description and results of our study of grouse nesting ecology. Chapter 4 follows with results from our efforts to better understand the ecology, survival, and behavior of hens with their broods. Chapter 5 entails the overall analysis of ruffed grouse survival, and the factors that affect the probability that a bird hatched in May will actually live to breed. As a complement to Chapter 5, Chapter 6 contains details of predators of grouse and their effects in the central Appalachians.

The next set of chapters covers food habits and

habitat and space use by ruffed grouse. In Chapter 7, the reader will find what the birds eat, when they eat it, and how this affects their condition. Chapter 8 provides results of detailed analyses of the sites selected for roosting by individual grouse and by broods. In Chapter 9 we present the habitat requirements of ruffed grouse in the Appalachians, and point out the substantial differences between their habitat requirements compared to the core range. Complementing the information on habitat requirements, we provide information on home range characteristics of ruffed grouse (Chapter 10) and their dispersal and tendency to reuse the same areas (Chapter 11).

We conclude the book with several chapters that build on the basic ecology summarized in Chapters 3 through 11. In Chapter 12, we provide guidelines for harvest management of ruffed grouse, and how we can maintain sustainable populations of birds to harvest. You can't have a population without habitat, and in Chapter 13 we present means by which habitat can be managed to enhance ruffed grouse populations. And, finally, Chapters 14 and 15 address management on private lands and the outlook for the future of ruffed grouse in the Appalachians. In the appendices we provide additional information on the references to support our work, lists of common and scientific names of animals and plants mentioned in the text, and additional information that may be of interest to some readers.

Throughout the book we have tried to present the "big picture" overview of the results of our study, striving to keep the important components of our work intact, while not burying the reader with excessive statistical and analytical detail found in the theses, dissertations, and scientific publications that have resulted from our efforts.

A study such as this on ruffed grouse (or any species, for that matter), with the level of cooperation and coordination among multiple agencies, states, and universities that we had over a six-year period on multiple sites is unparalleled. At times, the efforts to move the work forward were challenging, tiring, rewarding, and fun. We have enjoyed pooling our efforts to tell the tale of the ecology of the Appalachian ruffed grouse. We hope you enjoy the story.

DEAN F. STAUFFER, *Editor*

JOHN W. EDWARDS,
WILLIAM M. GIULIANO,
and GARY W. NORMAN,
Associate Editors

Acknowledgments

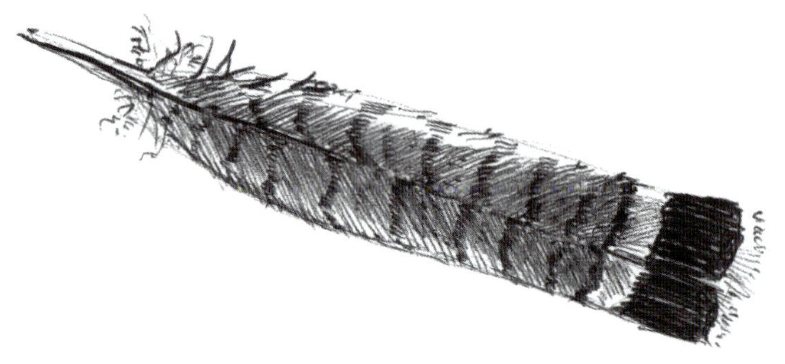

A work of this magnitude cannot be successfully undertaken without the support of many. There is a long list of organizations and individuals who contributed to our efforts to bring this project and book to fruition.

Tom Allen, Steve Bittner, Dan Dessecker, Mark Ford, Pat Keyser, John Organ, and Dave Samuel provided encouragement and guidance at the inception of the ACGRP.

Significant financial support was provided by the Richard King Mellon Foundation and we thank Michael Watson for his support throughout the project.

Each state cooperating on the project provided substantial financial and in-kind logistical support; we gratefully acknowledge such contributions from the North Carolina Wildlife Resources Commission, Virginia Department of Game and Inland Fisheries (Pittman–Robertson Federal Aid in Wildlife Restoration Project WE-99-R), Kentucky Department of Fish and Wildlife Resources, West Virginia Division of Natural Resources (Pittman–Robertson Federal Aid in Wildlife Restoration Project W-48-R), Maryland Department of Natural Resources (Pittman–Robertson Federal Aid in Wildlife Restoration Project W-61-R), Ohio Division of Wildlife and Rhode Island Department of Fish and Wildlife (Pittman–Robertson Federal Aid in Wildlife Restoration Project W-23-R).

Partial funding and additional logistical support for the project was provided by the Ruffed Grouse Society, USFWS Region V Northeast Administrative Funds, George Washington and Jefferson National Forest, MeadWestvaco Corporation, Champlain Foundation, North Carolina Wildlife Resources Commission, Pennsylvania Department of Conservation and Natural Resources, Pennsylvania Game Commission, Coweeta Hydrologic Lab, and the Campfire Conservation Fund.

The graduate students conducting much of the research received support from their institutions and we recognize the contributions of University of Tennessee Department of Forestry,

Wildlife and Fisheries; West Virginia University Wildlife and Fisheries Program, in the Division of Forestry and Natural Resources; Department of Fisheries and Wildlife Sciences at Virginia Tech; California University of Pennsylvania; University of Rhode Island; Fordham University; and Eastern Kentucky University.

Keith Warnke of the Wisconsin DNR, Al Stewart of the Michigan DNR, and Rick Horton of the Minnesota DNR were helpful with the collection of birds for comparison to grouse collected from ACGRP study sites, and their respective agencies provided the personnel to assist with collections.

Scott Klopfer assisted with GIS coverage for every study site.

Gary White provided assistance and training in program MARK used in the survival analyses.

Efforts in each cooperating state were enhanced by a diversity of people. We appreciate the contributions by the following people on study sites in each state:

Rhode Island: Scott McWilliams.

Pennsylvania: Amber Ausmus, Rich Ciaffoni, Michael Cooper, Steve DeSimone, Ross Garlopow, Jody Gregory, Mark Grietzer, Adrienne Leppold, Wayne Sech, and Lindsey Ware.

Ohio: Eric McAfee, Shane Berry, Lloyd Culbertson, Adam Duff, Jim Inglis, Lisa Irvine, Jason Lang, Jane Masklowski, Rick Smith, Bob Stoll, Jeremy Twigg, and Jim Yoder.

Kentucky: M. Shane Berry, Jennifer Kross, Jason Russell, Brandon Scurlock, Elizabeth Vincent, and C. Todd Williams.

West Virginia 1: J. Adams, T. Allen, C. Brown, J. Cromer, B. Cutwright, J. Evans, M. Eye, S. Gregory, R. Hartzell, S. Head, H. Jones, B. Knight, W. Lesser, J. Pack, G. Plaugher, D. Vandevander, C. Waggy, and S. Wilson

West Virginia 2: D. Arbogast, L. Berry, J. Craft, T. Dale, D. Gibson, J. Hajenga, E. Holland, S. Houchins, M. Hylton, B. Igo, C. Lawson, J. Morgan, R. Pettrey, E. Richmond, R. Roles, C. Ryan, G. Sharp, R. Sharp, R. Silvester, C. Taylor, G. Thorn, and S. Warner.

Maryland: Harry Spiker.

Virginia: David Steffen was instrumental in all parts of the project, and provided key guidance in the analyses of the population dynamics data. Bob Duncan and Bob Ellis provided key administrative support for the project. Denny Martin coordinated and flew the DGIF plane when birds were missing.

Virginia 1: Deerfield Ranger district, especially W. Buddy Chandler, and the Region 4 office of the Virginia Department of Game and Inland Fisheries (Al Bourgeois, Gary Spiers, Gene Sours, Nelson Lafon, John Pound, Kenny Sexton, and Rodger Propst). Technicians Chris Croson, Kevin Doherty, Matt Fields, Jim Inglis, Ellen Jedrey, Florence Kittleman, Jake Kubel, Jen Martin, Brad Pickens, Daly Sheldon, Kari Signor, Jennifer Steinbrecher, and Kris Stevens.

Virginia 2: Richard Clark, Scott Fluharty, Aaron Proctor, Ellen Jedry, Scott Johnson, Ross Garlapow, Todd Puckett, and Joe Secoges

Virginia 3: VDGIF personnel: John Baker, Jason Blevins, Milton (Punk) Bridgeman, Jonathan Chapman, Toby McClanahan, Mark Robinette, Dave Telesco, and Scott Whitcomb.

North Carolina: NC Wildlife Resources Commission personnel: David Allan, Brandon Allen, Jerry Anderson, Joffrey Brooks, Don Jones, Mike Parks, John Rogers, Mike Seamster, Gordon Warburton, and David Woody. US Forest Service personnel (Coweeta Hydrologic Lab): Wayne Swank and Jim Vose; (Bent Creek Experimental Forest): David Loftis; (Wayah Ranger District): Michael Wilkins. UT: Billy Minser and Monte Seehorn (USFS, retired), Steve Henson (Southern Appalachian Multiple Use Council), and George Taylor.

Illustration Credits

G. B. Bumann: pages 14, 17, 23, 31, 36, 41, 54, 64, 68, 70, 81, 87, 97, 114, 130, 159.

M. St.Germain: pages 11, 152, 156.

J. (Fettinger) Kleitch: page 125.

RUFFED GROUSE SOCIETY

1 Introduction

Dean F. Stauffer

The ruffed grouse is a popular gamebird throughout its range. Considered by some to be the "king of gamebirds," it is ardently sought by hunters wherever it occurs. In addition to the moniker of "ruffed grouse," this bird is also known in the Appalachians by a diversity of names, including moore fowl (SC), mountain cock (NC, SC), mountain partridge or mountain pheasant (VA, NC, SC), or by pheasant (throughout the area; Bump et al. 1947). Ruffed grouse are distributed from Alaska across central and southern Canada and the northern United States to the Atlantic Coast, south into the central Rocky Mountains and Appalachian Mountains. Its distribution coincides closely with that of big tooth and trembling aspen, except in the central Appalachians (Figure 1.1). Throughout most of the range of the ruffed grouse, aspen is considered a key component of ruffed grouse diet and cover. Exceptions to the aspen–grouse link occur in the Cascade Mountains in the west and in the central and southern Appalachian Mountains. The ruffed grouse is the most widely distributed resident gamebird in North America, while the aspens are one of the most widely spread tree species.

Ruffed grouse have been studied extensively over the years. However, most of the intensive research and studies conducted on grouse have taken place in the core of their range, where populations are relatively high. The first substantial study on ruffed grouse was conducted over an 11-year period, 1931–1942, in New York state by Gardiner Bump and his colleagues (Bump et al. 1947). Their effort represented a massive undertaking, and researched all aspects of the natural history, management, and propagation.

In the 1950s, Bill Marshall initiated studies on ruffed grouse on the Cloquet Forest in Minnesota. He was soon joined by Dr. Gordon Gullion. Their work, conducted in aspen-dominated habitats in Minnesota, has served as the basis for many of the management recommendations that have been

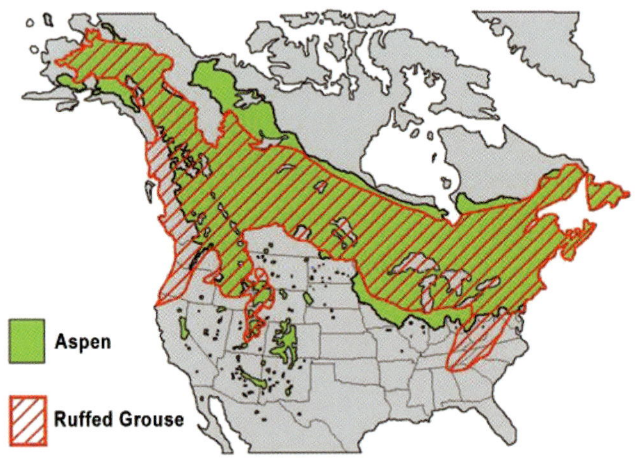

Figure 1.1

Distribution of aspen and ruffed grouse across North America.

developed for grouse throughout their range (Gullion 1965, 1970, 1977, 1983, 1984a,b; Gullion and Marshall 1968).

Additional work in Alberta (Doerr et al. 1974, Rusch and Keith 1971), Wisconsin (Dorney 1963, Kubisiak 1985, Destefano and Rusch 1986), and Michigan (Clark 2000, Larson et al. 2001) has contributed substantially to our understanding of all aspects of the ecology of ruffed grouse in boreal and northern hardwood forests where aspen is a dominant tree species, providing cover and food for the bird. In addition to the link to aspen where these studies were done, ruffed grouse populations throughout their northern range are subject to a 10-year cycle in abundance.

South of central Pennsylvania, things change quite a bit for ruffed grouse. Moving south along the Appalachians, we leave forests where aspen is a strong component, and enter woodlands that are dominated by oaks and hickories, with only small remnants of vegetation more typical of northern forests at the higher elevations in the mountains. In this region, ruffed grouse have a more patchy distribution and are less abundant than in the core of their range. Additionally, there is no evidence that these populations follow the 10-year cycle observed farther north. A relatively modest amount of research had been done on ruffed grouse in the central Appalachian region previously. Stewart (1956) studied grouse broods in Virginia, and Ralph Dimmick and his students conducted studies of ruffed grouse in Tennessee (Hollifield and Dimmick 1995, Kalla and Dimmick 1995). Hale et al. (1982) reported the results of a study on drumming log selection by ruffed grouse in Georgia. Perhaps the greatest amount of ruffed grouse research in the central Appalachian region was conducted previously by Roy Kirkpatrick and his students at Virginia Tech. They primarily conducted work on the nutritional ecology and food habits of ruffed grouse in this region (Norman and Kirkpatrick 1984, Servello and Kirkpatrick 1987, Hewitt and Kirkpatrick 1997a,b).

The work of these relatively few researchers indicated that the ecological relations of ruffed grouse may be different in this region on the edge of their geographic distribution. For example, the nutritional ecology of grouse is substantially different here. In the core area of their range, grouse rely heavily on aspen buds, twigs, and catkins to meet their nutritional needs for a substantial period of the year. In contrast, ruffed grouse diets in the central Appalachian region consist of leaves and seeds of herbaceous plants, acorns and buds of beech, birch, and cherry trees, and fruits of greenbrier, grape, and numerous other soft mast producers. Diets of grouse in the central Appalachian region tend to be higher in tannin and phenol levels, which are chemical compounds in some plants that serve as potential toxins. Additionally, diets of Appalachian grouse tend to have lower protein levels than found in the northern United States and Canada. The poor nutritional quality of grouse diets in the central Appalachian region may mean that grouse have to forage longer to attain adequate nutrition, which in turn may increase their risk of being taken by a predator. Additionally, the low nutrition may decrease body condition, reproductive potential, and chick survival.

Although there are relatively fewer grouse in the Appalachians, and their ecology is different than elsewhere, they are still an ardently sought-after gamebird. However, there are not as many grouse hunters in the central Appalachians as in the core states of their distribution. Estimated numbers of ruffed grouse hunters in 2001 were

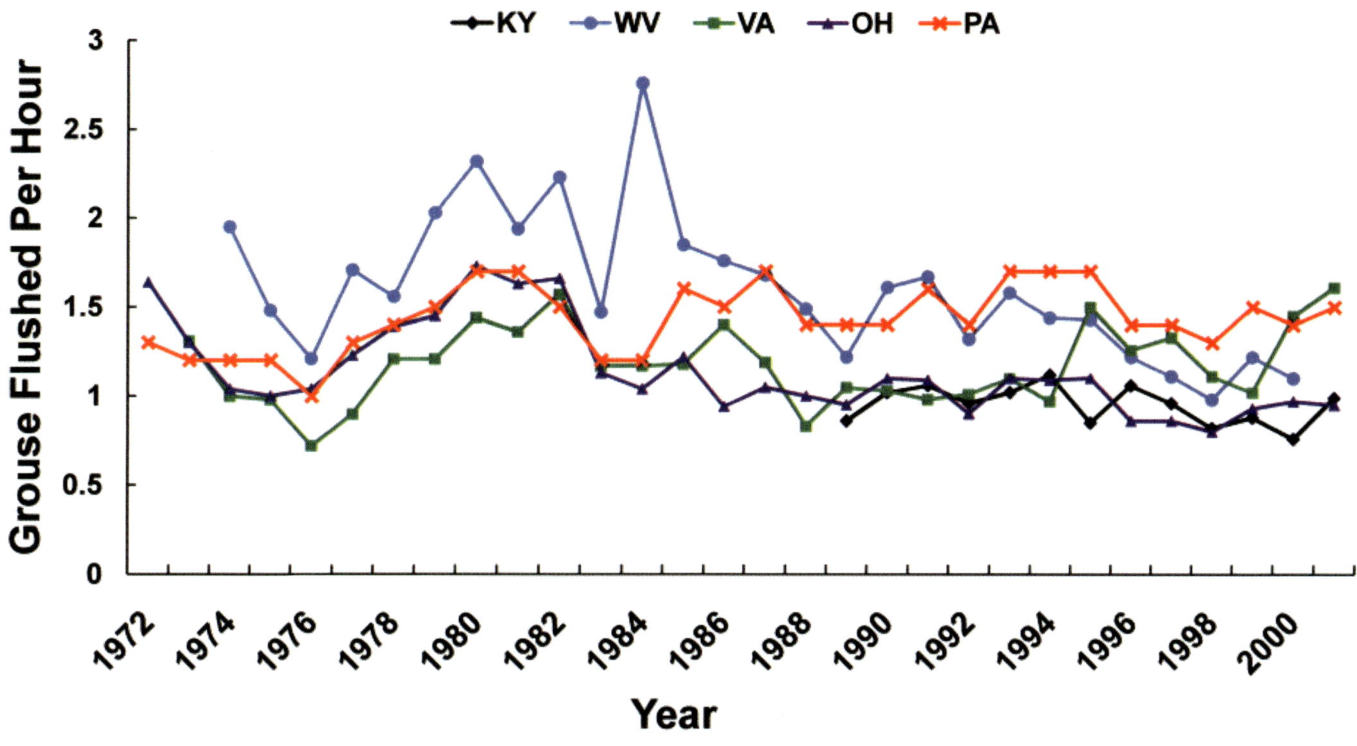

Figure 1.2

Trends in hunter-reported ruffed grouse flush rates for states within the ACGRP study area, 1972–2001.

101,000 for Minnesota, 124,000 in Wisconsin, and 116,000 in Michigan. In contrast, hunter numbers are relatively low, and declining, in the Southeast. In 1983, it was estimated there were 18,800 grouse hunters in North Carolina, 37,000 in Virginia, and 6,700 in Maryland. In 1989, the estimates were 9,400 in North Carolina, 7,000 in Maryland, and 24,000 in Virginia. These numbers declined to 2,500 hunters in Maryland and 19,000 hunters pursuing grouse in Virginia in 2001. Although the numbers of grouse hunters are low in this region, they are just as devoted in pursuit of their quarry as the more numerous hunters in the core range.

From 1972 to 2002, rates of grouse flushed per hunter hour varied, but were consistently low in Kentucky, West Virginia, Ohio, Pennsylvania, and Virginia (Figure 1.2). During the same period, total harvest declined in Maryland, Virginia, and North Carolina (Figure 1.3). Throughout this region, hunting seasons vary somewhat. Typically, seasons open in the middle of October, and continue until mid to late February. Thus, seasons tend to be longer in this region than in the Lake States. Over the past four decades, ruffed grouse populations have tended to decline throughout the central Appalachian region. The U.S. Fish and Wildlife Service initiated a program called the Breeding Bird Survey (BBS) in 1966. This survey was initiated to monitor breeding populations of birds throughout the United States and southern Canada. A summary of BBS data collected for ruffed grouse indicates that where grouse occur in Kentucky, Ohio, Pennsylvania, West Virginia, Maryland, Virginia, Tennessee, and North Carolina that populations have declined over the last 40 years, often at a rate of over 1.5% per year. Drumming data suggest that Virginia breeding populations have declined at a 3.1% rate per year from 1996 through 2005 (Norman 2006). Some of the decline may represent a reduction in early suc-

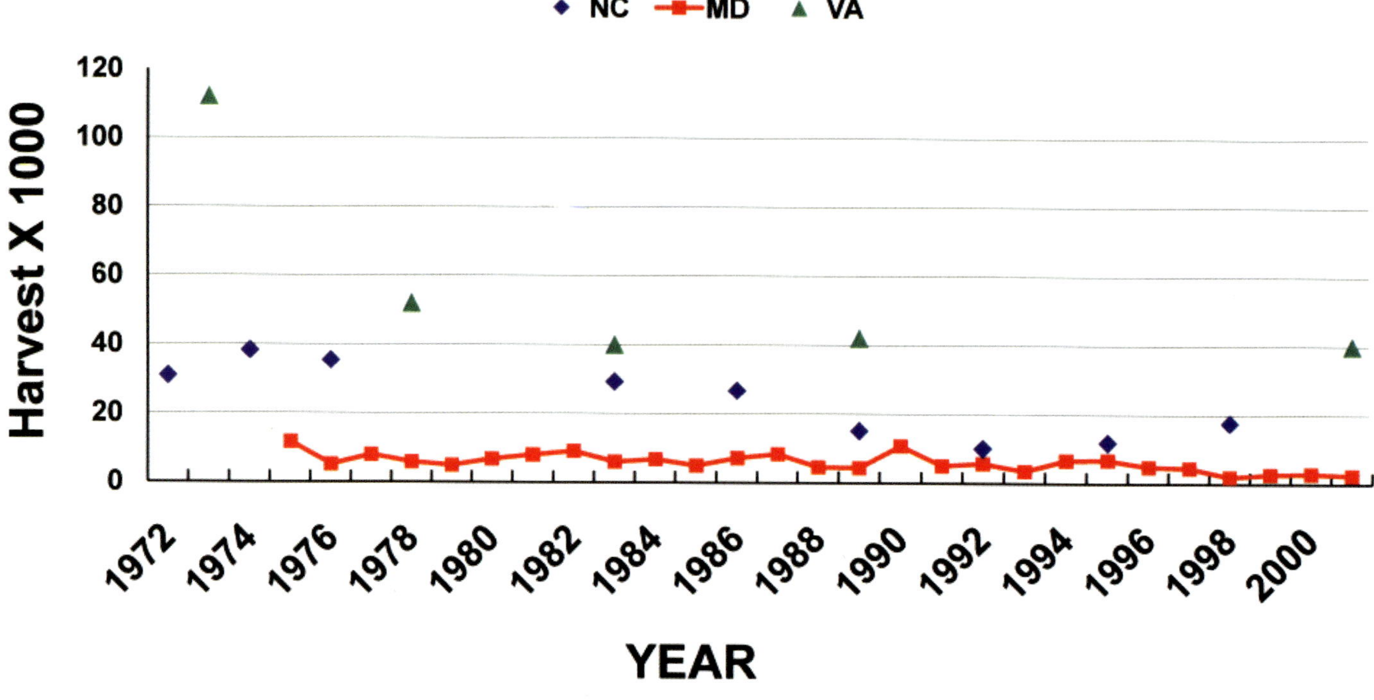

Figure 1.3

Trends in harvest rate for ruffed grouse in North Carolina, Virginia, and Maryland, 1972–2001.

cessional habitats preferred by grouse throughout the region, as forest harvest has declined and clear cuts have matured. The only exception to the trend in the region is at the far southern portion of grouse range in far southwest North Carolina, where populations have increased.

So, things were not looking good for ruffed grouse in the region. Populations were declining. Hunter numbers were down. Harvests had declined. Flush rates were relatively low and variable, which may reflect the fact that the lower number of hunters could still locate good coverts where grouse can be found, and work them to generate constant flush rates, even though overall populations were down. Because seasons tend to extend relatively late in the winter, some concern was expressed that late-season hunting might be having a negative effect. The worry was that birds harvested in January or February might not actually represent a "harvestable surplus," which is usually assumed for upland gamebirds. Rather, there was concern that the late-season harvest was adding to the normal winter mortality, and might be reducing the potential breeding stock heading into the spring nesting season. Prior research to determine whether hunting mortality adds to overall mortality in ruffed grouse, and is more than just taking the harvestable surplus, has not clearly answered the question. For example, Kubisiak (1984) found that an average harvest mortality rate of 44% added to overall mortality in Wisconsin. Bump et al. (1947) concluded that losses to hunting were not reducing overall populations when harvest was below approximately 50% in New York. And, in North Carolina, Monschein (1974) concluded that harvest mortality in ruffed grouse was taking only the surplus birds and not impacting the overall population. Thus, it is clear that the effects of grouse harvest, especially late-season harvest, on overall grouse survival were not yet well understood.

In fall of 1995, the biennial meeting of the

Southern Grouse Workshop was held in southwest Virginia. A number of biologists working with grouse throughout the several-state region attended this workshop, along with some invited grouse researchers from the Lake States. A number of issues concerning the perceived decline in grouse numbers and late-season harvest were discussed, and it was clear that there was a need to conduct some well-designed research to assess the overall status of grouse in the central Appalachians. At that time, there was an ongoing study of ruffed grouse survival being conducted in Ohio by the Ohio Department of Natural Resources, and a new study had just gotten underway in West Virginia to look at habitat selection on lands subject to active forest management, conducted by the WV Department of Natural Resources in conjunction with West Virginia University. Additionally, a new project was initiated by the Virginia Department of Game and Inland Fisheries in cooperation with Virginia Tech to study grouse survival and the potential effect of late-season harvest.

Using these initial studies as incentive, a meeting was held in Elkins, WV in February 1996 to discuss the possibility of coordinating research efforts. Attendees at the meeting included Dave Swanson from OH DNR, Tom Allen from the WV DNR, Gary Norman of VA DGIF, and Dean Stauffer from Virginia Tech. Discussions were held to explore how we might be able to coordinate the data collection in a consistent manner, which would allow combining some analyses to evaluate patterns in grouse ecology across Ohio, West Virginia, and Virginia. These discussions led to the idea of holding a larger scoping meeting, inviting biologists and managers from the larger region, to explore the possibility of developing a region-wide grouse research effort.

Thus, in May of 1996, a meeting was held at Fort Lewis Lodge in Virginia to discuss the next steps. There were representatives from game/wildlife agencies of Ohio, West Virginia, Virginia, Pennsylvania, Maryland, and Kentucky. Additionally, we had representation from the USFWS Federal Aid program, the Ruffed Grouse Society, West Virginia University, and Virginia Tech. Two days of fruitful discussion led to the formal creation of the Appalachian Cooperative Grouse Research Project. Initially, we had cooperators and study areas in Kentucky (one site), Ohio (two sites), West Virginia (two sites), Maryland (one site), and Virginia (three sites). The original objectives of the ACGRP were:

1. Determine the survival and cause-specific mortality rates of ruffed grouse;

2. Determine productivity — how many females nest, how many eggs are laid, and how many hatch and survive — and estimate how many newly hatched birds join the population each year;

3. Determine whether hunting mortality adds to overall mortality in grouse, or if it just takes the "surplus" birds; and,

4. Develop computer models of population growth that could be used as tools to guide population management for ruffed grouse.

We established a series of cooperators' meetings to develop consistent protocols for data collection across all the sites. These meetings were initially held at two-month intervals as we designed the approaches used to capture grouse, track and monitor their status, and collect habitat and ancillary data. After about two years, as the methods became standardized, we reduced the meeting frequency to twice a year to discuss progress and status of various states' projects. In 1998, the Richard King Mellon Foundation provided substantial support to the ACGRP. With their support came a new study site in Pennsylvania. In 1999, Ohio had completed their intended work, and ceased collecting data that were contributed to the project. Also in 1999, we were joined by researchers from the University of Tennessee who had a study site in North Carolina, and a site in Rhode Island was added.

Our first two years of data on survival indicated that hunting mortality might in fact be reducing overall survival, and depressing total population levels. Thus, as the cooperators evaluated this preliminary evidence, we developed a proposal for a second phase of the study, which would entail the closure of harvest on some study sites for several years to allow us to compare survival rates in grouse populations subjected to hunter harvest to those populations that are not harvested. In 1999, we entered Phase II of the study. We were able to secure cooperation of the relevant state wildlife agencies, and closed the hunting season on three study sites with the highest harvest rates for the first three years of the study (see Table 2.1 for a list of the sites that were closed to hunting). Survival of grouse on these three sites was then compared to survival on four sites that were open to harvest for the full six years of the study.

The success of the ACGRP was dependent on the support and cooperation of a wide diversity of individuals and organizations. We have previously acknowledged individuals and organizations that contributed to the effort. Table 1.1 lists the major players in this unique undertaking. Having multiple cooperators and multiple study sites with similar data collection techniques allowed us to leverage the support from any individual state or site to secure and maintain funding. Each state was able to contribute perhaps 10–20% of the total cost of the work, yet they gained access to 100% of the results. This represents a 500–1000% return on the investment. The cooperative nature of the work, and pooling data from across multiple study sites and making those data available to researchers and graduate students, allowed us to generate information and uncover ecological patterns and processes that would not have been obvious from one or a few of the study sites. We had the four stated objectives of the study that were met by collection of the data from all the sites. However, the graduate students involved with the work also developed subprojects in addition to the primary objectives to investigate other aspects of ruffed grouse ecology. Some of the additional investigation topics included brood habitat use and survival, landscape-level analyses of habitat and home range size, roost site selection and habitat analysis, effects of prescribed burning on grouse habitat, and dispersal and movements of grouse. Results from these additional projects are presented in the subsequent chapters.

Table 1.1. Major cooperators and contributors to the Appalachian Cooperative Grouse Research Project.

AGENCIES	Kentucky Department of Fish & Wildlife Resources Maryland Department of Natural Resources Ohio Department of Natural Resources Rhode Island Department of Environmental Management U.S. Fish and Wildlife Service U.S. Forest Service Virginia Department of Game and Inland Fisheries West Virginia Division of Natural Resources
UNIVERSITIES	California University of Pennsylvania (2 students) Eastern Kentucky University (1 student) Fordham University (1 student) Frostburg State University (1 student) The Ohio State University (1 student) University of Tennessee (3 students) Virginia Tech (6 students) West Virginia University (4 students)
OTHER COOPERATORS	Pennsylvania Game Commission MeadWestvaco Corporation Richard King Mellon Foundation The Ruffed Grouse Society

2

Study Areas and Field Methods

Dean F. Stauffer

Study Areas

We studied ruffed grouse populations on 12 sites in eight states throughout the central Appalachian region (Table 2.1; Figure 2.1). Land ownership varied across sites and included National Forest land, state public land, and industrial land owned by MeadWestvaco Corporation. Study sites ranged in size from 2,000–11,000 hectares (4900–26,950 acres). The proportion of forest age-classes (seedling, sapling, pole, and sawtimber) varied across sites due to differences in past timber management activities. Current management ranged from no active harvest to selective harvest and clearcutting. MeadWestvaco lands had the most active timber harvesting programs and thus the greatest proportion of sapling-age stands. Hunting seasons on all study areas typically ran from early October to late February, with daily bag limits ranging from one to four grouse and possession limits of four to eight.

Study sites (except OH1 and OH2) were clas-

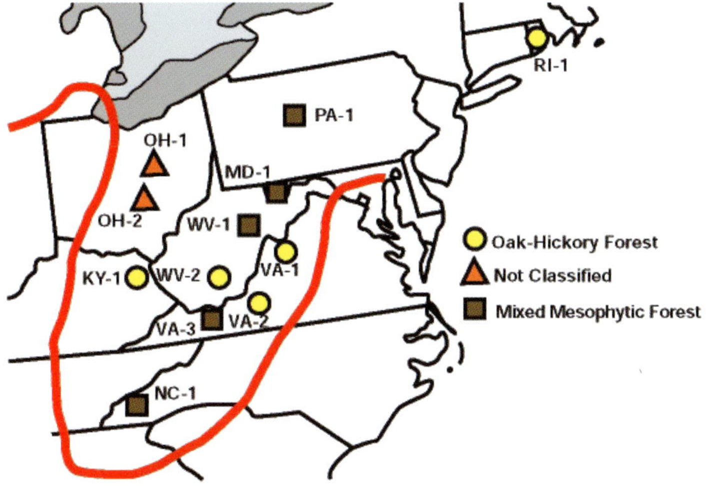

Figure 2.1

Location of Appalachian Cooperative Grouse Research Project study sites, 1996–2002. The solid line indicates the approximate distribution of ruffed grouse in eastern North America.

Table 2.1. Description of study sites participating in the Appalachian Cooperative Grouse Research Project, 1996–2002.

Study Area	Ownership	Counties	RPI[a]	Forest Type	Hunting Treatment[b]	Years
KY1	State	Lawrence	8.21	Oak-Hickory	Closed	1996–2002
MD1	State	Garrett	33.62	Mixed-Mesophytic	Open	1996–2002
NC1	Federal	Macon	32.4	Mixed-Mesophytic	N/A	1999–2002
OH1	State, Private	Athens, Vinton, Meigs	N/A	N/A	N/A	1996–1999
OH2	State, Private	Coshocton	N/A	N/A	N/A	1996–1999
PA1	State	Clearfield, Elk	35.96	Mixed-Mesophytic	N/A	1998–2002
RI1	State	Kent	25.54	Oak-Hickory	N/A	1999–2002
VA1	Federal	Augusta	25.0	Oak-Hickory	Open	1997–2002
VA2	MeadWestvaco	Botetourt	27.81	Oak-Hickory	Open	1996–2002
VA3	State	Smyth, Washington	33.13	Mixed-Mesophytic	Closed	1996–2002
WV1	MeadWestvaco	Randolph	34.73	Mixed-Mesophytic	Open	1996–2002
WV2	MeadWestvaco	Greenbrier	28.15	Oak-Hickory	Closed	1996–2002

[a]RPI = relative phenology index.
[b]Hunting treatment refers to hunting experiment during last 3 years of the project.

sified as either oak-hickory or mixed-mesophytic forest associations based on literature review, species composition of canopy trees, and tree abundance data collected as part of the ACGRP (J. Tirpak, Fordham University, unpublished data) and a Relative Phenology Index (RPI, Table 2.1). The RPI estimates the timing of phenological events and duration of growing seasons based on latitude, longitude, and elevation according to what is called Hopkins' bioclimatic rule (Hopkins 1938). We calculated RPI values for each site based on the mean latitude, longitude, and elevation of ruffed grouse radio-telemetry locations averaged across years. The RPI values calculated for each study site indicated that growing seasons on mixed-mesophytic sites (i.e., higher RPI values) were shorter than on oak-hickory sites despite the interspersion of the two forest associations in the Appalachian region (Figure 2.1). The OH1 and OH2 study areas were not classified due to lack of canopy tree composition and abundance data.

Oak-hickory forests were dominated by chest-

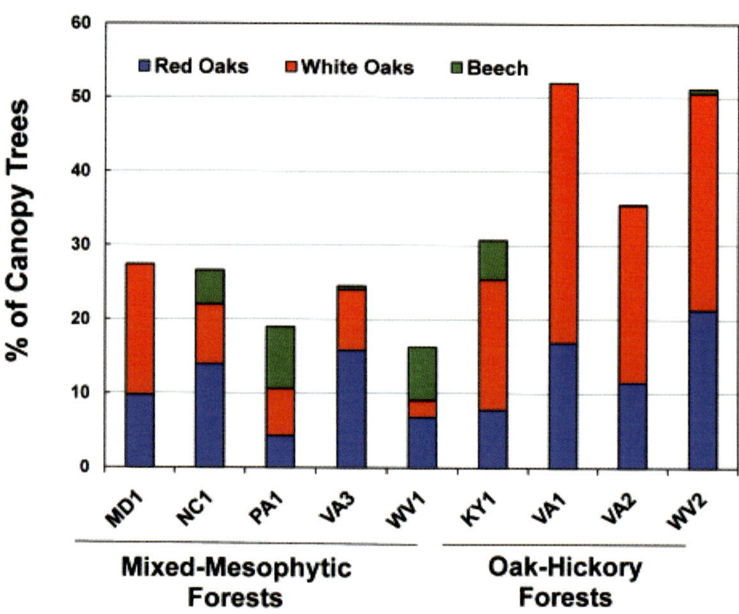

Figure 2.2

Percentage of canopy trees on ACGRP study sites represented by members of the red and white oak groups and American beech trees. Data were collected at randomly located 0.04 ha plots. The percentages were derived from a total of 50,701 trees sampled across the study areas 1996–2002.

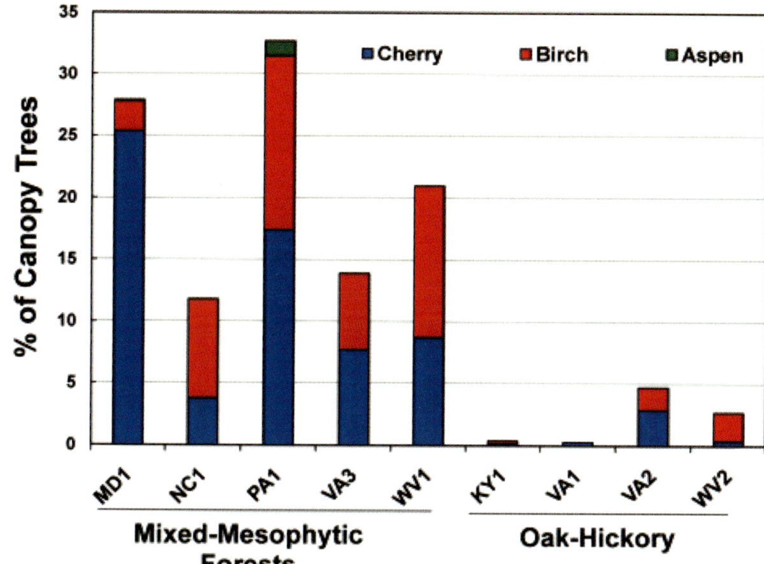

Figure 2.3

Percentage of canopy trees on ACGRP study sites represented by aspen, birch, and cherry trees. Data were collected at randomly located 0.04 ha plots. The percentages were derived from a total of 50,701 trees sampled across the study areas 1996–2002.

nut, white, northern red, scarlet and black oaks, and shagbark, pignut, mockernut, and bitternut hickories. Other important tree species were red, sugar, and striped maple; American beech; table mountain, white, Virginia, and pitch pine; and eastern hemlock. Mountain laurel and great rhododendron were important understory species. Dominant canopy species on mixed-mesophytic sites were sugar maple, red maple, yellow birch, basswood, black and pin cherry, yellow-poplar, white pine, American beech, northern red oak, and eastern hemlock. Other important species were white ash, white oak, and aspen. Hard mast producing species, including oaks and beech, were present on mixed-mesophytic and oak-hickory forests, but more abundant on oak-hickory sites (Figure 2.2). Aspen, birch, and cherry were more abundant on mixed-mesophytic sites than on oak-hickory sites (Figure 2.3).

Field Methods

To gather the data necessary to meet the objectives of the ACGRP, it was necessary to capture grouse and fit them with radio transmitters. Radio-telemetry technology provides researchers opportunities to obtain information about an animal's daily and seasonal activities that would otherwise be quite difficult to determine. The decision to use radio-telemetry is not a trivial one; it can be relatively expensive in terms of equipment and time necessary to capture and monitor animals. However, for the ACGRP, implementing a radio-telemetry study was well worth the time and effort, and critical to our achieving our objectives. Here we describe the general methods used to capture and follow the birds; additional details of methods used in particular aspects of the study are presented in the appropriate chapters.

ACGRP personnel trapped ruffed grouse from August to December (fall) and February to April (spring) between 1996 and 2002 in lily-pad traps as described by Gullion (1965). The traps consisted of a 15–20-meter (50–66-ft) drift fence constructed from chicken wire that connected two trap bodies (Figure 2.4). Grouse walking through the cover would encounter the drift fence, and then follow along the fence until they walked into the trap. Traps were placed in locations where we had previously seen grouse, and at sites that appeared to be suitable habitat. During the trapping season we checked traps each morning and evening.

When a bird was captured, we removed it from the trap and placed it into a handling bag (Figure 2.5) for processing. We recorded the weight of each bird and determined age and sex based on feather

To Find a Grouse

Once we had captured the grouse, placed leg bands and radio-transmitters on them and released them, we had to track them. Putting a radio transmitter on an animal allows the researcher to follow it during what we assume to be its normal activities. We attempted to relocate every grouse with a transmitter twice a week. The approach taken was to travel to established listening stations throughout the study areas, and use the telemetry receivers and antennas and determine the compass bearing or azimuth for the antenna direction in which the strongest signal was heard. We attempted to get azimuths from at least three different listening stations. To illustrate the process, assume that we had the following area with five telemetry stations as indicated by the numbered hexagons. The star symbol indicates the location of a bird with a transmitter.

Now, assume we first go to telemetry station #1 on the trail up to Short Mountain. We go to the station, and are able to pick up the signal from the bird, we scan the horizon and determine the direction of the peak signal, and we then record the azimuth indicated by the line from station #1 as seen above.

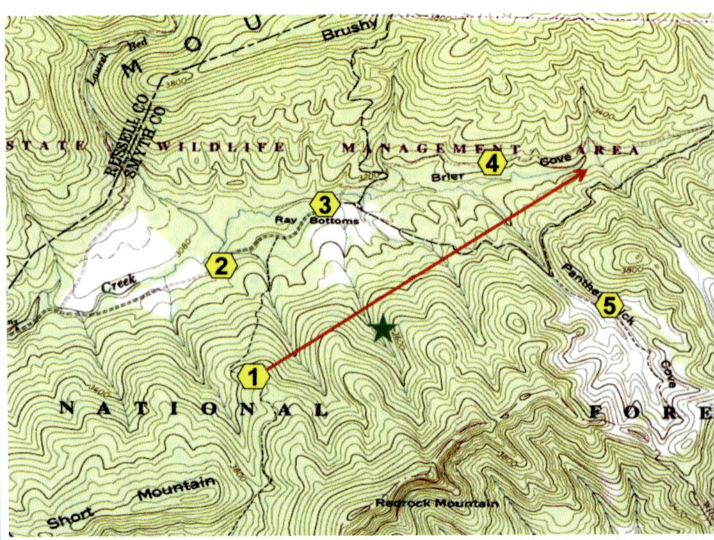

Now, this first azimuth shows a couple of things. First, there is error associated with the azimuths taken on birds with transmitters. Our abilities to detect the peak signal, the terrain, weather, and a host of other factors influence the error with telemetry systems. We often can come pretty close to the true azimuth, but numerous studies have shown that typical error is around plus/minus 5–6 degrees, meaning that an estimated azimuth of 90 degrees would be anywhere within a range of 84–96 degrees. The second thing that is obvious is that one observation is not enough! Even if this azimuth were exactly accurate, we would not know where the bird was along that line; thus, we must take at least two azimuths.

Assume that we next travel to telemetry station #3 and take our next azimuth, resulting in the following:

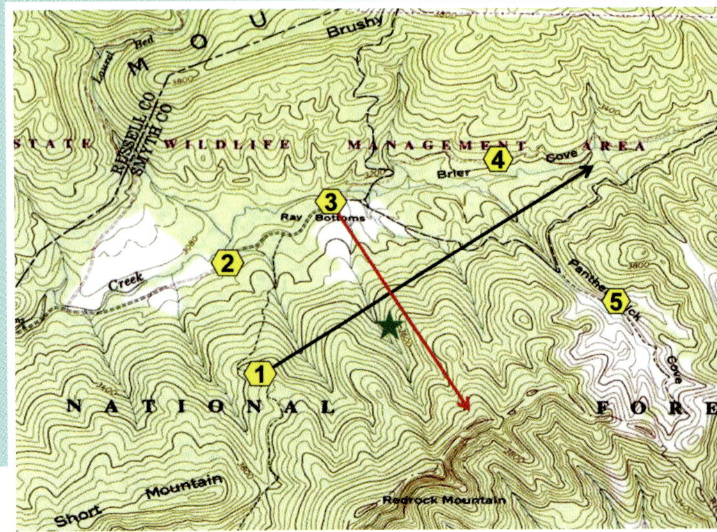

This second azimuth also has slight error, but it does give us two crossing lines. In the past, much telemetry was conducted with only two azimuths, with the investigators assuming that the animal was where the lines crossed. Sometimes this may be correct, but more likely than not, the assumed location indicated by the intersection of two azimuths may not be the true location of the animal. Research has shown that more azimuths give a more accurate and reliable location.

So, let's assume we travel along to station #5 in Panther Lick Cove. We take an azimuth from the station there and get the following result:

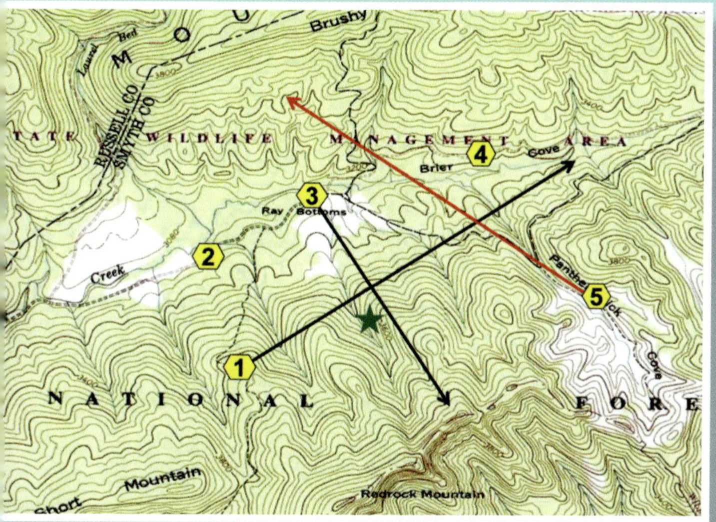

Whoops! What is going on here? Given the first two azimuths, we might have expected the next azimuth to be somewhat close to the intersection of the first two. This third one illustrates a common challenge with telemetry in mountainous situations, that of "signal bounce." In this case, because the bird is on a hillside just above a drainage, and is buffered from station #5 by a couple of ridges, the stronger signal comes from a location to the north of the true location, and is actually a reflected signal coming from the mountainside to the north of where the bird is located.

So, we need more data. Our dedicated field technician then goes to stations #2 and #4 to acquire additional azimuths and gets the following results:

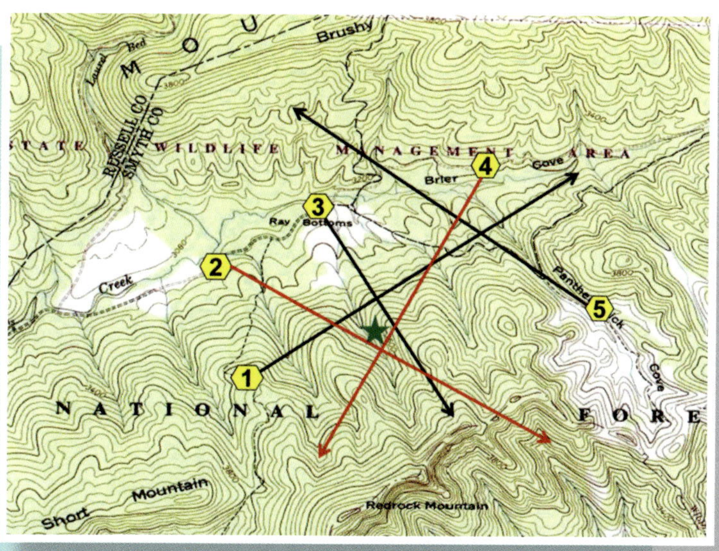

Now we have enough information to estimate where the bird actually was located. We first would eliminate the azimuth from station #5 — it clearly is an outlier and not accurate. Using the remaining four observations, we could simply select a location within the area delineated by the azimuths, and this is what often was done in the past. However, better ways to analyze these types of data have been developed. The majority of radio-telemetry studies now use what is known as the Lenth Estimator to analyze the azimuths and assign a location. This sophisticated mathematical approach was initially developed in the 1970s as a means to allow rescue parties to home in on emergency signals such as those that might come from a downed plane or lost hiker/skier. Enterprising wildlife scientists recognized that search and rescue is similar to the issue of locating the exact location of a wild animal with a transmitter. This approach analyzes the intersections of all the azimuths, the distance from the receiving stations, the error associated with each azimuth, and then estimates the statistically most likely location of the transmitter. For all of our analyses that used grouse locations, we used the Lenth Estimator to determine the birds' locations.

—Dean F. Stauffer

Figure 2.4

Photograph of the trapping setup. Left: A drift fence connecting two lily-pad traps used to catch ruffed grouse on the ACGRP study sites. Right: a bird caught in a trap. Grouse moving through the cover would encounter the drift fence, and then walk along the fence and into the trap. This design takes advantage of the grouse's tendency to walk around obstacles rather than simply flying over them. *Photos: ACGRP Files.*

characteristics as described by Davis (1969) and Kalla and Dimmick (1995). Birds were fitted with a uniquely numbered, aluminum leg band and a 10-gram necklace-style radio transmitter (Figure 2.5). Each radio was designed with a motion-detection switch that increased the pulse rate of the transmitter's signal after being motionless for eight hours. Thus, when checking bird status, if the pulse rate of the signal from a particular bird had doubled, we assumed the bird had died, and we would locate it and attempt to determine the cause of death. The transmitters we used had a guaranteed life of 12 months, and many were still on the air up to 20 months or more. In "typical" Appalachian habitats, we could detect the transmitters up to about 1.6 kilometer (a mile) away.

Figure 2.5

A ruffed grouse in a handling bag (left), having been weighed, sexed, and measured, and fitted with the radio collar (center), and a grouse being released (right) and ready to join the study. *Photos: Dean F. Stauffer (left); Robert Long (center and right).*

Each radio transmitter and leg band had contact information and an offer of a $25 reward. This helped to ensure that when a bird was harvested by a hunter, or found dead, that we would be notified of the fact.

There is the possibility that capture, handling, and putting the band and radio transmitter on a bird may affect its behavior and survival after release. Thus, we allowed each bird a seven-day "acclimation" period before we considered them to be part of the study. After seven days, we assumed they had adapted to their transmitter and leg band, and that their behavior represented that of a typical grouse. We attempted to capture at least 40 grouse each year on each site to provide enough data for a meaningful analysis.

We attempted to determine the status of each bird at least twice a week for the life of the radio transmitter or of the bird (Figure 2.6). We used triangulation methods to determine the location of each bird (White and Garrott 1990). Triangulation entailed acquiring at least three (in some cases two) or more azimuths on the transmitter signal from known locations on the study site. We used Global Positioning System (GPS) units to determine the exact location of the listening stations. We attempted to gather the azimuths in less than 30 minutes to minimize the effect of bird movement on the estimated location. We used a computer program to analyze the azimuths, or plotted azimuths on a topographic map, to estimate the most likely location of each bird. We then used these locations to analyze home ranges, movement, and habitat selection.

By the end of 1999 we had telemetry information on approximately 800 birds. We had preliminary information that showed that reproduction and chick survival were critical components of grouse ecology. Thus, to more efficiently make use of our time and effort, beginning in 2000 we concentrated on tracking females in the spring so we could locate all nests, and we only monitored whether males were dead or alive, but did not take the extra azimuths necessary to estimate location.

We captured a total of 3,118 ruffed grouse between fall 1996 and spring 2002; this includes 413 birds that were captured twice. The average trap rate was 2.4 grouse/100 trap-nights (Table 2.2). Trap success was greater for traps set near forest stand edges compared to traps set in mature forest stands, which reflects the general habitat selection of grouse. Trap success varied quite a bit among our study sites. In Pennsylvania, we captured, on average, six birds for every 100 trap-nights of effort. The study sites in Ohio and West Virginia also provided relatively good success, with a capture rate of at least three birds for every 100 trap-nights. We were not as successful on other areas; for example, on the North Carolina site, and the VA1 site in Virginia, our capture rate was less than one bird per 100 trap nights. In addition to birds trapped, we also noted birds flushed adjacent to

Figure 2.6

Conducting radio telemetry in the field to determine the location and status of one of the ACGRP grouse fitted with a radio transmitter. *Photo: Jennifer Kleitch*

Table 2.2. Summary of ruffed grouse fall trapping success in the Appalachian region for each study site, 1996–2002. Flushes were birds seen at the trap site, but not caught.

Study Area	*n*	Grouse /100 Trap Nights	Flushes /100 Trap Nights
		Mean	Mean
KY1	6	1.41	0.60
MD1	5	2.17	1.81
NC1	3	0.89	
OH1	1	3.20	1.03
OH2	2	4.59	1.66
PA1	4	6.00	1.98
RI1	3	1.23	0.51
VA1	5	0.87	2.22
VA2	6	1.06	1.27
VA3	6	1.13	0.35
WV1	6	3.00	2.13
WV2	6	4.71	

Table 2.3. Summary of age ratios of ruffed grouse captured in the fall in the Appalachian region for each study site, 1996–2002.

Study Area	*n*	Juvenile : Adult Female	Juvenile Female : Adult Female
		Mean	Mean
KY1	6	0.53	0.30
MD1	6	1.31	0.70
NC1	3	0.53	0.32
OH1	3	0.45	0.27
OH2	4	0.36	0.19
PA1	4	0.74	0.38
RI1	3	0.47	0.18
VA1	5	1.03	0.44
VA2	6	0.24	0.13
VA3	6	0.28	0.17
WV1	6	0.32	0.12
WV2	6	0.42	0.15

the traps (Table 2.2). While flush rates were lower than capture rates, study areas that had higher capture rates also had higher flush rates.

The overall ratio of captured juvenile grouse to adult females was 0.56:1.0 (Table 2.3). As with the capture rates, there was quite a bit of variation in these ratios. We recorded a high of 1.31:1 on the Maryland site, compared of a low of 0.24 juveniles per adult female on the VA2 site. The sex ratio was slightly skewed and averaged about 1.48 males for every female, overall (Table 2.4). The sex ratio of all birds captured ranged from nearly even (1.04:1) on the North Carolina site, to more than two males for each female in Rhode Island and the WV2 study area.

Table 2.4. Summary of sex ratios of ruffed grouse juveniles and adults captured in the fall in the Appalachian region for each study site, 1996–2002.

Study Area	*n*	Male : Female	
		Mean	95% C I
KY1	6	1.77	0.99 – 2.55
MD1	6	1.19	0.75 – 1.63
NC1	3	1.04	0.64 – 1.45
OH1	3	1.44	0.72 – 2.15
OH2	4	1.32	1.06 – 1.57
PA1	4	1.25	0.79 – 1.70
RI1	3	2.66	1.59 – 3.74
VA1	5	1.76	0.61 – 2.91
VA2	6	1.22	0.92 – 1.52
VA3	6	1.32	0.88 – 1.77
WV1	6	1.70	1.35 – 2.06
WV2	6	2.14	0.68 – 3.59

3 Nesting Ecology

John M. Tirpak,
Patrick K. Devers,
William M. Giuliano,
John W. Edwards,
Benjamin C. Jones,
and Craig A. Harper

Across most of their range, ruffed grouse are a relatively common forest gamebird associated with early successional aspen habitats. In the Appalachians, though, grouse have historically occurred in low numbers at the periphery of their range where aspen is uncommon or lacking. This regional difference in population density may be related to forest composition through its effect on the quantity and quality of available forage resources (Servello and Kirkpatrick 1988). In northern regions, aspen buds represent an abundant high-energy winter food that grouse are well adapted to exploit. Alternatively, in the Appalachians, birds depend on inconsistently available mast (e.g., grapes, acorns, beechnuts) and low quality evergreen vegetation (e.g., mountain laurel leaves) to survive the winter (Norman and Kirkpatrick 1984). This dietary difference manifests in southern grouse populations as poorer condition of hens entering the breeding season, which may have important consequences on productivity (Beckerton and Middleton 1982). Because grouse rely on high annual recruitment of young to replenish losses from high annual mortality (Sharp 1963, Gullion 1970), low productivity can result in low population numbers.

Ruffed grouse in the Appalachians have been experiencing declines not seen in the northern range. Although believed to be linked to the loss of high quality early successional habitat (Dessecker and McAuley 2001), the mechanism of these declines is again most likely reduced productivity. Hence, the already lower productivity of Appalachian grouse populations may be further depressed by diminishing habitat quality.

Nest success and brood survival are the two most important components of reproduction in ruffed grouse populations (Tirpak et al. 2005). Therefore, management prescriptions that target these vital population characteristics may have the greatest impact on grouse populations. Although many researchers consider brood survival the more significant of these factors, its importance may be overstated because brood size is often underestimated (Godfrey 1975a). Alternatively, nest success may be more important than commonly believed because nest survival is usually overestimated due to a bias towards locating successful nests (Larson et al. 2003). Furthermore, in the central and southern Appalachians, some nest predators (e.g., snakes) are more common than in northern regions, which may elevate the role nest success plays in the dynamics of Appalachian grouse populations.

Because nest site habitat factors influence nest success in many gamebirds, managers may be able to augment grouse productivity by increasing the availability of high-quality nesting habitat. Unfortunately, descriptions of grouse nesting habits are mostly anecdotal and quantitative data on nests and nesting success in the Appalachians are lacking. Therefore, the ACGRP took special efforts to measure reproductive effort in Appalachian grouse populations and to identify the factors associated with reproductive success. Specifically, we determined: (1) the earliest, latest, and average dates of nest initiation and hatching, (2) the proportion of hens that nested, (3) average clutch size and egg hatchability, (4) nest success rates, causes of nest failure, and identity of specific nest predators, (5) renesting rates, and (6) habitat selection and the influence of habitat factors on these rates at the nest site, study area, and regional scales.

Figure 3.1

Typical ruffed grouse nest. *Photo ACGRP Files*

Methods

Grouse nests are typically simple depressions on the ground in which hens lay their eggs (Figure 3.1). Because there is no structure to the nest itself and hens allow a close approach prior to flushing (some won't even flush when touched), systematic searches for grouse nests typically produce poor results. Therefore, the ACGRP relied solely on telemetry to locate nests of radio-tagged hens.

During April and May, cooperators intensively (at least three times per week) monitored hens for signs of nesting. A bird was suspected of incubating when it was repeatedly found at the same location. Researchers would then home in on the bird and flush it to locate the nest and count the eggs. Because grouse do not normally abandon their nests when flushed during incubation, this method results in the most accurate estimates of clutch size for the maximum number of nests with the minimum impact on incubating hens. Once nests were found, researchers conducted all subsequent monitoring remotely via telemetry to minimize disturbance. When a hen was found repeatedly away from her nest site, a second visit to the nest was made to determine if she was successful and how many eggs had hatched. Researchers relied on eggshell evidence to determine if a nest was successful or not. Eggs that hatch are typically cracked in a neat circle along the large end of the egg as a result of the chick "pipping" its way into the world (Figure 3.2). Hens do not remove the eggshells after hatching, so they are typically still in the nest bowl when the brood leaves it just a few short hours after hatching. Eggs depredated by predators were invariably in fragments or altogether absent (Figure 3.2). We considered nests successful if at least one egg hatched and calculated nest success as the percentage of successful nests. Because nests were typically located shortly after initiation of incubation, we assumed this nest success rate accurately reflected the overall success rate among nests (Mayfield 1961). Nests destroyed during egg laying were not included in these calculations; however, these losses were considered minimal and hens that lose nests during this early stage are likely to initiate another clutch

Figure 3.2

Top: Successful ruffed grouse nest with hatched eggs.
Bottom: A failed nest showing depredated eggs.
Photos: ACGRP Files

Figure 3.3

Nest camera setup used to monitor nesting ruffed grouse.
Photo: Brian W. Smith.

(Johnsgard and Maxson 1989). We calculated egg hatchability as the number of eggs that hatched from those laid in ultimately successful nests. Again, losses of eggs during the laying stage from ultimately successful nests were assumed to be minimal. Birds that nested unsuccessfully were intensively monitored for renesting until July first of each year. Similar protocols were followed for re-nests as for first nest attempts. Clutch size, nest success, and egg hatchability were calculated independently from first nests. We assumed birds that did not nest between the first of April and July first did not breed, and calculated breeding probability as the proportion of hens that attempted to nest at least once.

At a subset of nests at the WV1 site, researchers installed infrared cameras to continuously monitor hen activity and identify specific nest predators. These cameras were mounted so only the lens and the attachment arm were located near the nest to minimize the effect of monitoring on nest concealment. A cable connected the camera lens to a video recorder and power source located about 20 meters from the nest site, which allowed researchers to change video tapes and batteries daily without flushing incubating hens (Figure 3.3).

In 0.04-hectare (0.1-acre) circles around each nest site, we measured a suite of habitat characteristics to quantify vegetation structure: percent tree canopy, ground vegetation, and coarse woody debris cover, small (<8 cm dbh) stem density (stems/ha), and basal area of trees (m^2/ha). In addition, we classified slope, distance to the nearest road or opening, timber size class, midstory vegetation volume, and understory type for each plot (Table 3.1). As part of a complementary study, we also collected data on these habitat variables along a systematic array of sampling points on five sites (MD1, PA1, VA2, VA3, and WV1) to characterize available habitat for use in assessing habitat selection.

Table 3.1. Habitat characteristics recorded at nest sites and randomly located sites 1996–2002. We also show the habitat characteristics for successful and unsuccessful nests. Habitat features that are substantially different between nest and systematic, or successful and unsuccessful samples can be considered important in influencing nest site selection, or success.

	Habitat Selection[a]		Nest Fate[b]	
Habitat Variable	**Nest**	**Systematic**	**Successful**	**Unsuccessful**
Basal area (m^2/ha)	24.7	19.7	27.8	23.4
Deciduous canopy cover (%)	80.8	83.5	81.9	73.7
Coniferous canopy cover (%)	5.1	6.5	6.2	4.7
Ground cover (%)	47.2	49.8	46.7	53.2
Coarse woody debris cover (%)	17.3	9.2	25.2	19.0
Stems/ha	13,382	10,701	9918	8916
PERCENT OF OBSERVATIONS IN EACH CATEGORY				
Distance to road or opening				
Close (<10 m)	54.8	17.2	65.0	54.0
Moderate (11–100 m)	20.5	74.5	17.1	14.0
Far (>100 m)	24.7	8.3	17.9	32.0
Timber size class (%)				
Sapling (<12.5 cm dbh)	34.2	38.2	36.8	42.0
Pole (12.5-27.8 cm dbh)	37.0	35.6	29.1	30.0
Sawtimber (>27.8 cm dbh)	28.8	26.3	34.2	28.0
Midstory volume (%)				
Open (<20%)	28.8	3.5	20.5	28.0
Moderate (20–50%)	67.1	90.0	72.7	60.0
Closed (>50%)	4.1	6.5	6.8	12.0
Understory volume (%)				
Open (<20% woody and <30% herbaceous)	49.3	43.5	40.2	50.0
Herb (>30% herbaceous)	31.5	37.5	35.0	32.0
Wood (>20% woody)	19.2	19.0	24.8	18.0
Slope (%)				
Moderate (11–30%)	47.9	44.8	53.0	62.0
Steep (>30%)	8.2	6.1	17.9	10.0

[a]Based on 73 nest and paired random samples from 5 study areas.
[b]Based on 167 nests (117 successful and 50 unsuccessful nests) from 8 study areas.

To evaluate nest habitat selection, we compared habitat at each nest to all systematic points located within 291 meters of it, the radius of an average winter–spring home range of female grouse on these areas (26.6 ha; Whitaker 2003). Only nests located on the five systematically sampled study areas were used to assess nest site selection. Alternatively, we compared successful and unsuccessful nests from eight sites (KY1, VA1, and WV2 in addition to the five above) to determine the potential influence of habitat on nest success.

Results and Discussion

On most sites, hens began incubation in mid April to early May, generally starting earliest on the southernmost sites (Table 3.2). Given that a grouse will on average lay two eggs every three days (Maxson 1977), these dates correspond to the first eggs being laid in early April on most sites. Although there was some yearly variation in nest initiation dates, the range of dates was generally within two weeks of the mean. The range of dates was narrower on the southern sites than the northern areas, likely a result of more variable weather conditions on these latter sites (late-season snow or a delayed spring). Incubation averages 24 days for grouse and average hatch dates closely followed this time frame on most sites (23–27 days after initiation), with most first broods hatching in mid to late May. Hatch dates for re-nests ranged from mid to late June. All hens were off their nests by July 1.

Nesting rates were consistently high across most sites, with every hen attempting a nest on six of the nine sites (Table 3.3). Even on the three sites where some hens failed to nest, nesting rate was high (at least 89%) and non-nesting hens were only observed in a single year. Failure to nest likely reflects poor condition of hens entering the breeding season. The three sites with imperfect nesting rates were all dominated by oak-hickory forest. Birds on these sites were reliant upon the hard mast crop produced by a few tree species. Not surprisingly, years where some birds failed to nest typically followed falls that had a poor mast crop. Conversely, mixed-mesophytic sites contained a greater diversity of plant species and grouse on these sites had a broader forage base and a number of alternative foods when mast crops failed. Hens on these sites apparently were able to gather enough energy to nest in each year. Age of the hen did not appear to influence nesting rate, as eight of the 15 birds that failed to nest were adults.

Table 3.2. Initiation and hatching dates for ruffed grouse nests in the Appalachians, 1996–2002.

	Incubation Initiation Date		Hatch Date	
Study area	Range	Mean	Range	Mean
KY1	4/14 – 5/6	4/21	5/8 – 5/30	5/16
MD1	4/2 – 5/26	5/3	5/15 – 6/19	5/30
NC1	4/21 – 5/3	4/28	5/15 – 5/27	5/22
PA1	4/24 – 5/25	5/7	5/18 – 6/18	6/1
VA1	4/19 – 5/4	4/27	5/13 – 5/28	5/21
VA2	4/17 – 5/3	4/25	5/12 – 5/27	5/19
VA3	4/19 – 5/10	4/30	5/17 – 6/3	5/25
WV1	4/12 – 5/24	5/1	5/6 – 6/19	5/25
WV2	4/18 – 5/7	4/29	5/12 – 5/28	5/22

Table 3.3. Nesting rates for ruffed grouse nests in the Appalachians, 1996–2002. Numbers in parentheses represent number of hens or nests observed / total nests or hens.

		First nests				Re-nests			
Study Area	Years	Breeding probability	Clutch size	Success rate	Egg hatchability	Breeding probability	Clutch size	Success rate	Egg hatchability
KY1	6	100% (19/19)	9.0	63% (12/19)	82% (79/96)	43% (3/7)	7.5	66% (2/3)	100% (15/15)
MD1	6	100% (36/36)	11.1	64% (23/36)	78% (181/232)	31% (4/13)	6.8	50% (2/4)	91% (10/11)
NC1	2	100% (16/16)	10.4	88% (14/16)	94% (137/145)	50% (1/2)	N/A	0% (0/1)	
PA1	4	100% (61/61)	9.8	56% (34/61)	84% (284/337)	30% (8/21)	6.7	50% (4/8)	96% (26/27)
VA1	4	89% (32/36)	9.5	84% (27/32)	84% (203/241)	0% (0/5)	N/A	N/A	N/A
VA2	6	96% (47/49)	8.9	48% (22/46)	91% (181/200)	0% (0/24)	N/A	N/A	N/A
VA3	6	100% (36/36)	10.1	75% (27/36)	89% (237/265)	0% (0/9)	N/A	N/A	N/A
WV1	6	100% (92/92)	10.2	72% (63/88)	94% (548/583)	27% (6/22)	7.6	50% (3/6)	100% (26/26)
WV2	5	89% (55/62)	9.1	64% 27/42)	81% (129/159)	0% (0/15)	N/A	N/A	N/A

Overall, on the nine sites where nest data were collected, researchers followed 407 hens and located 394 nests (Table 3.3). Clutch size was at least eight eggs on each site, typical for a species that suffers high annual mortality, but lower than that recorded in northern regions (11.5 in New York and 11.0 in Wisconsin; Rusch et al. 2000). Because grouse chicks are precocial and leave the nest at hatching, each egg requires a significant investment of energy from the female. At approximately 20 grams, each egg weighs about 4% of the female's total weight, and a large clutch may represent up to 50% of the hen's total weight. Therefore, like nesting rate, average clutch size may partly be a function of hen condition; however, total clutch size is likely limited the by physical ability of the hen to incubate the eggs. The lower nesting rate in the Appalachians corroborates this hypothesis, as does the variation in average clutch size among the areas in this study. Again, forest composition, which has an influence on the consistency of food resources from year to year, appears to be an important determinant of clutch size; birds on the four oak-hickory sites averaged lower clutch sizes than birds on the five mixed-mesophytic sites (9.1 vs. 10.3 eggs, respectively).

Researchers were able to accurately determine fate for 376 of the 394 nests located during the course of this study. Of these 376 nests, 249 were successful (66%) across all sites (Table 3.3), a rate similar to the 63% success rate observed by Bump et al. (1947) in the only other large scale study, in New York state, that investigated nesting success. Success rate varied among sites (48–88%); however, unlike nesting and clutch size, differences among study areas did not reflect forest composition. Instead, nest success is likely a function of the abundance and diversity of the local predator community, as has been observed for other birds (Rodewald and Yahner 2001). Nest success did differ dramatically among years; again, possibly as a result of annual changes in predator and other prey abundance or the sensitivity of the small number of nests located each year to random events.

On the WV1 site, researchers equipped 25 nests with cameras and documented 23 different nest visitors (Dobony et al. 2003). However, only five visits resulted in egg depredation, with a black rat snake, a black bear, a long-tailed weasel, and two raccoons positively identified as nest predators. One additional monitored nest was depredated, but the camera failed to record the event. However, researchers observed a raccoon at the nest site when they returned to change the battery in the morning and presumed this animal was responsible for destroying the nest. Another camera documented two unsuccessful attempts by a long-tailed weasel to capture an incubating female on consecutive nights. Although the female flushed from the nest both times, the weasel did not destroy any of the eggs nor did the hen abandon the nest. Five eastern chipmunks and a short-tailed shrew were also observed at nests, but none removed eggs or impacted nest fate (all eggs subsequently hatched).

Egg hatchability was lower in this study (87.6%; Table 3.3) than that recorded for other parts of the grouse range (96.5%; Bump et al. 1947). Egg infertility was rare, but some eggs contained embryos too weak to make it out of the egg. The inability of these chicks to break out of the egg may be related to a lower investment of energy in these eggs by females that enter the breeding season in poorer condition. However, this was relatively uncommon and hens may reduce the number of eggs they lay rather than lower the quality of individual eggs. Egg hatchability was lower on oak-hickory sites (83.4%) than mixed-mesophytic (86.4%), but the response was not as pronounced as for other parameters. By far, the majority of eggs that did not hatch were eggs lost during the incubation stage. Whether these were individually depredated or were broken and removed by the hen is unknown. Foxes and small mammals may repeatedly visit nests and remove single eggs (Bump et al. 1947), and the abundance of these predators in the Appalachians may be a factor in this higher loss rate.

Like most parameters associated with nesting, the renesting rate of grouse in the Appalachians (17.7%; Table 3.3) was lower than in more northern areas (56–67%; Small et al. 1996, Larson et al. 2003). Again, within the Appalachians, this rate varied with forest type and was lower on oak-hickory (5.9%) than mixed-mesophytic (26.0%) sites. Although clutch size and success rate were lower for re-nests than first nests, sample sizes were generally too low to draw any definitive conclusions. However, lower clutch sizes in second nests would be expected due to the lower energy reserves available to the hen for egg production after she has produced her first clutch. Similarly, higher predation rates for second nests are not surprising in light of the higher activity and greater abundance of some nest predators later in the summer. Higher predation rates during the late spring and summer may explain in part why grouse nest so early in the spring. Higher egg hatchability rates in renests is puzzling, but this rate may not be statistically meaningful because of the relatively few (11) successful re-nests in this study.

Nest Habitat

Grouse do not construct elaborate nests. Nests are typically situated against the base of a tree, stump, or boulder or against a log or in a brush pile. These nest substrates may serve as "backstops" to prevent ambush from behind by predators (Bergerud and Gratson 1988) while the hen is on the nest and vulnerable. Use of these microsites was reflected in the habitat structure selected by females for nest sites. Basal area (which is related to the number of trees) was about 25% higher and coarse woody debris loads (a measure of stumps and logs) were about 85% greater at nest sites than randomly sampled locations that represent what is available for the grouse to select from (Table 3.1). Therefore, nest habitat selection in grouse may be driven primarily by the availability of suitable nest sites.

Basal area may be a better indicator of suitable nest habitat than coarse woody debris; others have found that grouse more commonly nest against large trees than logs or stumps (at about a 2:1 ratio; Bump et al. 1947, Larson et al. 2003). This pattern may also reflect the forest age used for nesting. Although grouse are most common in early successional sapling stands during the winter, they typically nest in pole and sawtimber stands (Maxson 1978). Use of these timber size classes has been attributed to the preference of grouse for nesting in open understories where birds can easily detect predators (Bump et al. 1947, Gullion 1977). However, ground cover did not differ between nest locations and habitat that was available to them. Instead, grouse may nest in stands with larger diameter trees simply because of the greater availability of suitable nest sites in these stands.

Where large trees are not available, coarse woody debris may be an important nest substrate. Nesting in sapling stands was most common on the PA1 site (24 of 29 nests), where a tornado created a large swath of early successional habitat with abundant coarse woody debris on the ground. Conversely, sapling stands on other study areas were the result of logging operations that removed most of the large-diameter coarse woody debris. These stands were rarely used for nesting (9 of 49 nests). Where coarse woody debris is abundant, sapling stands may be as attractive to nesting grouse as pole or sawtimber stands with high basal area and larger trees.

At the stand level, basal area was also the most important predictor of nest success, followed by coarse woody debris, deciduous canopy cover, ground cover, and distance to a road or opening. Nest success increased with basal area, coarse woody debris, and deciduous canopy cover, and decreased with increasing ground cover and in locations >100 m from a road or opening (Table 3.1). If the probability of nest predation is related to the number of potential nests sites a predator must search (Siepielski et al. 2001), as the number of large trees or logs around a nest increases, the likelihood of a predator searching the particular tree or log where a nest is located decreases and the nest is more likely to be successful. This likely explains why basal area and coarse woody debris were so strongly associated with nest habitat even though a single large tree or log would suffice as a nest site.

Greater deciduous canopy cover and reduced ground cover were also associated with successful nests. A dense midstory of deciduous cover is often cited as an important component of nesting habitat because it reduces sunlight reaching the forest floor and prevents the growth of dense understory vegetation (Gullion 1977), which may hinder predator detection (Bergerud and Gratson 1988). However, we may not have observed selection for these stand characteristics because hens may not be capable of fully assessing canopy or understory conditions during incubation when the first eggs are laid prior to bud break and the emergence of forbs in the spring. Because grouse are exceptionally well camouflaged and nest on the ground, however, they may not need to rely on understory vegetation for concealment from potential predators and will nest in a variety of understory conditions.

Surprisingly, grouse nest success was lower farther from a road or opening. Nesting along edges is thought to reduce nest success by increasing exposure to predators that are more abundant in these habitats (Gates and Gysel 1978). Distance to edge habitat also did not affect predation of artificial grouse nests in an intensively managed forest in central Pennsylvania (Yahner et al. 1993), nor did it affect grouse nest success in New York (Bump et al. 1947). Grouse are a disturbance-dependent species and are likely adapted to nesting near edges. Their excellent camouflage, high nest attentiveness, and brazenness when confronted by predators all likely reduce predation and may contribute to why these birds can be more successful in edge areas than other species. Because canopy gaps are where the first green vegetation appears in spring, incubating females often feed along these areas. By nesting near edges, hens may be able to further reduce their foraging times and time away from the nest when both hens and their nests are more vulnerable to predation.

At the larger, landscape level, habitat surrounding nests also influenced nest success (except on the North Carolina study area). However, the relations were typically weak and were not consistent across sites. On the Virginia sites, higher percentages of 10–20-year-old forest were related to increased nest success, while the opposite was true on the West Virginia 2 site. Older forest was associated with higher nest success on the Pennsylvania site, where increasing oak and moister habitats around nests also were related to improved nest success. On the West Virginia 2 site, mesic areas contained more successful nests than drier sites. Evergreen forest was negatively associated with nest success on the Maryland site.

Nest success was marginally affected by habitat features at the landscape scale, with older forests generally associated with increasing nest success. The weak relation of most habitat features to nest success may reflect the limited effect of habitat at this scale. Because grouse are sedentary during most of the incubation period (Maxson 1977), they may rely more on their camouflage and the immediate microhabitat characteristics surrounding nests for protection from predators than on favorable landscape compositions. However, the use of older forests is a pattern observed across most of the grouse range (Maxson 1978). Older forest stands support greater basal area, which provides an abundance of nest sites that grouse select for nesting.

Maximum ruffed grouse productivity is likely determined before the first egg is even laid by the condition of the hen entering the breeding season. Forest composition plays a critical role in this equation by governing the availability and abundance of potential forages. On sites dominated by only a few mast-producing species, managers should increase the diversity of the forage base to provide a buffer against mast failure. Planting vines, shrubs, or trees or encouraging their growth via forest management practices (reviewed in Chapter 13) are two methods to achieve this goal. Additionally, identifying the strongest mast producers and improving their yield may result in more consistent crop and potentially more stable productivity and numbers in the grouse population.

Current grouse management in the central and southern Appalachians focuses mainly on creating early successional forest to provide winter cover for ruffed grouse. Although these efforts are critical, managing for secure nest habitats may also benefit grouse. Retaining pole and sawtimber stands in close proximity to sapling stands would increase the availability of safe nest sites in close proximity to the early successional habitats used in winter and for foraging. Alternatively, where residual basal area is reduced below 5.7 m^2/ha (25 ft^2/acre) managers should ensure that large diameter coarse woody debris (e.g., logging slash, felled snags, and stumps) cover exceeds 20% so an adequate number of potential nest sites are available in these stands. These recommendations are only appropriate when applied in conjunction with large-scale even-aged harvests, as early succession habitat is still necessary for ruffed grouse in the winter.

Why Statistics?

Benjamin Disraeli said, "There are three kinds of lies: lies, damn lies, and statistics." We sincerely hope that in this book we have no lies or damned lies, but we did find it necessary to use statistics. If we had been able to capture every individual grouse, count every tree, enumerate each and every drumming log — basically document every grouse movement and *everything* that was present on each study site — we would not need statistics at all. However, all the data we collected represent a sample of the larger areas; we did not capture and follow every grouse, we did not measure every square inch of habitat, and so on. Because our data represent a sample of the larger "whole," it is necessary to use statistical methods to summarize the limited sample we observed, make inferences about what it means for all grouse throughout the Appalachians, and to evaluate presumed relationships. In the course of analyzing and evaluating all the data collected on the ACGRP, we used a wide diversity of statistical methods to help identify meaningful ecological relationships and patterns in our sample of data. Although often dry and boring for the public and even wildlife managers, the theses, dissertations, and scientific publications that have resulted from the ACGRP efforts contain a multitude of statistical tests and models with the academically interesting descriptions of the science behind the results. The intent of this book is not to bore the reader, but to provide in a readable manner the results of our scientific research. The information we present, however, is fully supported and backed up by rigorous statistical analyses, even if the gory details are not reported here.

While minimizing reporting of statistical details in this book, we wish to emphasize the importance of employing proper statistical designs and analyses when conducting research such as that reported here. Appropriate statistical analyses allow us to detect and confirm real patterns and meaningful results in our data while eliminating mistaken conclusions about the data patterns we might see by chance. This process helps us to better understand the ecological relations of ruffed grouse to their environment.

For example, let's say that on one of our sites we found that 10% of home ranges on average were composed of early successional habitats. Just looking at that one number, we might assume that early successional habitat really isn't very important to grouse. However, this single estimate of use doesn't mean much in and of itself. Now, assume we had collected information on the composition of the habitat in the larger area that was available to the birds, and we found that only 3% of the area was early successional. Compared to the 10% early successional cover used by grouse within their home ranges, it would appear that grouse are selecting areas with more early successional cover available. But, because both the habitat use and availability estimates are based on samples of grouse and habitat observations, and not the whole area, a statistical test can help us to determine if this difference of seven percentage points (10%–3%) is large enough to be meaningful and not just the result of a chance outcome. Several different statistical methods might be used to evaluate this difference. If we were to find a ***statistically significant*** result, we could assume that this difference of seven percentage points is too large to happen just by random chance, and that the grouse are selecting home ranges with more early successional cover than what is available at random in the environment.

Throughout this book we will use terminology such as ***statistically significant***, when the reader encounters such wording, it will indicate that the results being presented or discussed were found to reflect "real" patterns, and were unlikely to arise randomly from the data.

Because the large majority of the data we collected represents samples, we would typically calculate the average or mean value (e.g., such as a tree density, bird weight) and report that. There is variation around this average value — some data points collected would be larger, and some would be smaller. To help indicate the confidence we have with our estimates of the mean, we often calculated and reported a statistic called the ***standard error***. A small standard error value relative to the mean value indicates that we have a relatively good estimate of the mean. We sometimes calculate a ***confidence interval*** as another way to show the level of confidence we have in our mean estimates. This interval generally is the mean value, plus or minus two standard errors. An interpretation of this interval is that if we were to select another sample of whatever it is we are measuring, we would expect that 95% of the time the mean value of this new sample would fall within this range.

—Dean F. Stauffer & David Steffen

4

Brood Ecology

Brian W. Smith, Benjamin C. Jones, John W. Edwards, Craig A. Harper, Patrick K. Devers, Chris Dobony, Scott Haulton, John M. Tirpak, Jennifer (Fettinger) Kleitch, William M. Giuliano, and David A. Buehler

Recruiting young birds into the breeding population is critical to sustain and increase ruffed grouse populations. However, ruffed grouse throughout their range face a hazardous period between hatching and their first breeding season. In fact, survival rates of young ruffed grouse are usually much lower than those of adults, and most mortality in young birds occurs during the first few weeks after hatch (Haulton 1999, Dobony 2000, Smith 2006). Early in the ACGRP planning stages, cooperators decided to study many aspects of ruffed grouse broods—their behaviors, habitat use, movements, causes of mortality, and survival rates—in order to help determine factors that might limit recruitment of young grouse in the Appalachians.

Despite the importance of the early stages of a ruffed grouse's life, there was little information available to provide a foundation for our work. Additionally, most previous studies were not conducted in the Appalachians, nor did they provide sufficient information for drafting management recommendations. A review of available studies showed that, compared with the entire range of ruffed grouse, reproductive success and recruitment were lower in southern portions of the ruffed grouse's range (Davis and Stoll 1973, Harris 1981, Kalla and Dimmick 1995). This led us to investigate whether low reproductive success and/or recruitment of young to the breeding population were contributing to low population densities in the Appalachians, and what factors (habitat, food availability, movements, weather, etc.) might influence chick survival throughout the region. In this chapter, we summarize the studies conducted under the ACGRP umbrella that addressed chick survival, habitat use, and movements.

Brood Survival and Causes of Chick Mortality

We collected two sets of information related to brood survival. As a part of the larger ACGRP, we monitored the survival of all hens with broods at three- and five-week intervals. Additionally, we conducted an intensive study of chick survival on three study sites.

Overall Chick Survival. As previously noted, it has been long suspected that ruffed grouse in the Appalachians have lower productivity and recruitment than grouse in the Lake States and southern Canada. Biologists also have suggested that low chick survival is an important factor contributing to relatively low abundance of ruffed grouse in the Appalachian region. The ACGRP was able to evaluate this suspicion—we monitored 467 females during nest and brood seasons from 1997–2002 in the Appalachians. We gathered a substantial amount of data from these females and broods, gaining insight into factors potentially limiting grouse populations in the region. We estimated chick survival to 35 days of age. We used our count of the number of eggs that hatched as the estimate of the initial brood size. We then located the hen and counted the number of chicks alive at 21 and 35 days of age. This information allowed us to derive an estimate of the proportion of chicks that survived for three and five weeks after hatching.

Chick survival in the Appalachian region was poor, to say the least. Overall, for the study duration, an average of only 22% of ruffed grouse chicks survived to 35 days of age (Figure 4.1; Norman et al. 2004). Norman et al. (2004) reported that chick survival to 35 days post-hatch was higher on mixed-mesophytic forests (39%) than on oak-hickory forests (21%). For comparison, Rusch and Keith (1971) found chick survival to 84 days (12 weeks) in Alberta to be over 50%. When compared to our results, it is obvious that chick survival in the Appalachians is markedly different from that in the northern portions of their range, but why? Hard mast may explain some of this difference. We found chick survival was positively correlated with hard-mast production the previous fall (Norman et al. 2004, Devers 2005). This provided additional evidence that ruffed grouse productivity and recruitment in the Appalachian region may be strongly influenced by the quality and availability of food resources, especially hard mast. Food availability and quality is important for successful reproduction in birds. Females must be able to efficiently find and consume high quality food to improve their body condition, egg quality, and

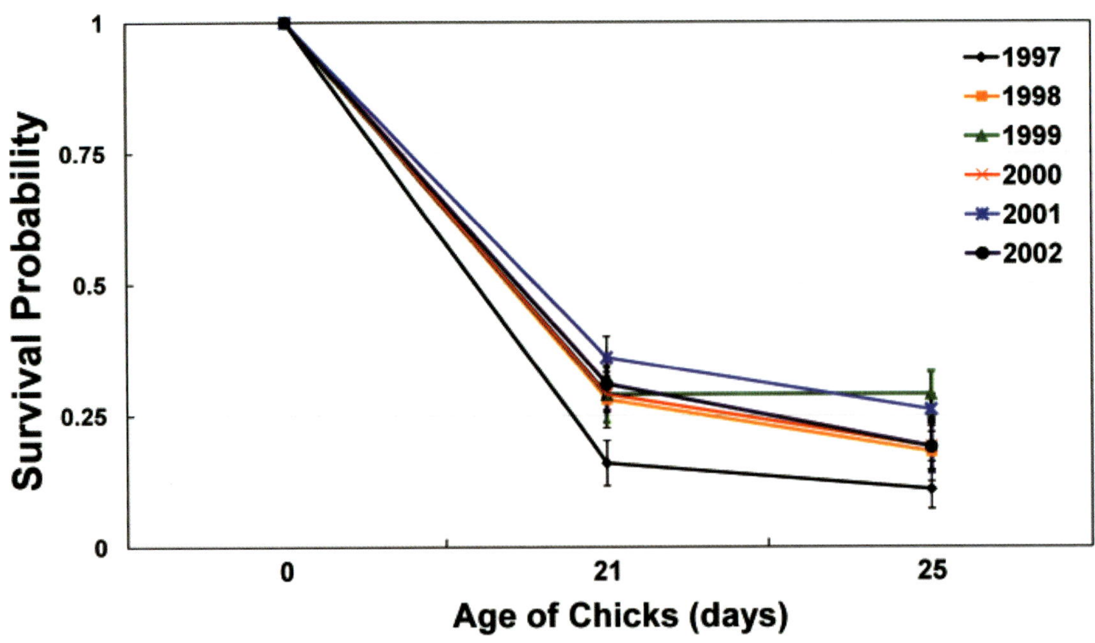

Figure 4.1

Ruffed grouse chick survival to 21- and 35-days post-hatch in the Appalachians, 1997–2002. Vertical bars represent the 95% confidence interval on the estimate.

Table 4.1. Average number of ruffed grouse chicks alive at weeks one, three and five. The average number of chicks per successful hen includes broods that lost all chicks, whereas the average number per surviving brood includes only broods with at least one living chick. Data were collected at eight sites in 1997 and nine sites in 1998 in the Appalachians.

		Successful Hen			Surviving Brood		
Year	Week	Broods[a]	n[b]	Chicks/ brood	Broods	n	Chicks/ brood
1997	1	26	68	2.6	17	68	4.0
	3	31	64	2.1	17	64	3.8
	5	30	34	1.1	13	34	2.6
1998	1	14	55	3.9	11	55	5.0
	3	18	49	2.7	12	49	4.1
	5	17	36	2.1	10	36	3.6
Years Combined	1	40	123	3.1	28	123	4.4
	3	49	112	2.3	29	112	3.9
	5	47	70	1.5	23	70	3.0

[a] number of broods
[b] number of chicks

chick survival. In years with poor mast production, females enter the reproductive season under stress and have less fat and protein reserves. If birds lack sufficient reserves, their eggs may have smaller yolks, which results in weaker chicks.

Intensive Chick Survival Study. We conducted an intensive survival study on chicks on three of the ACGRP sites (VA2, VA3, and WV2). We found survival rates were lowest the first week following hatch during 1997 and 1998 (Figure 4.2). Chick survival the first week post-hatch averaged 26–36%, and loss of complete broods was high (36–42%) during this time over the two-year period. Chick survival across all study sites in the Appalachians showed similar trends in survivorship at one, three, and five weeks of age (Table 4.1); sharpest declines in survivorship occurred during the first week, and then gradually decreased between weeks one and five post-hatch. Survivorship values at five weeks ranged between 0.11 and 0.13 for the two-year period, and survivorship to 13 (1997) and 10 (1998) weeks of age was 0.04 and 0.11, respectively (Figure 4.2). Hens that successfully hatched at least one egg had an average of fewer than two chicks at the end of five weeks, and the average surviving brood had three chicks, but this does not include broods that lost all their chicks (Table 4.1).

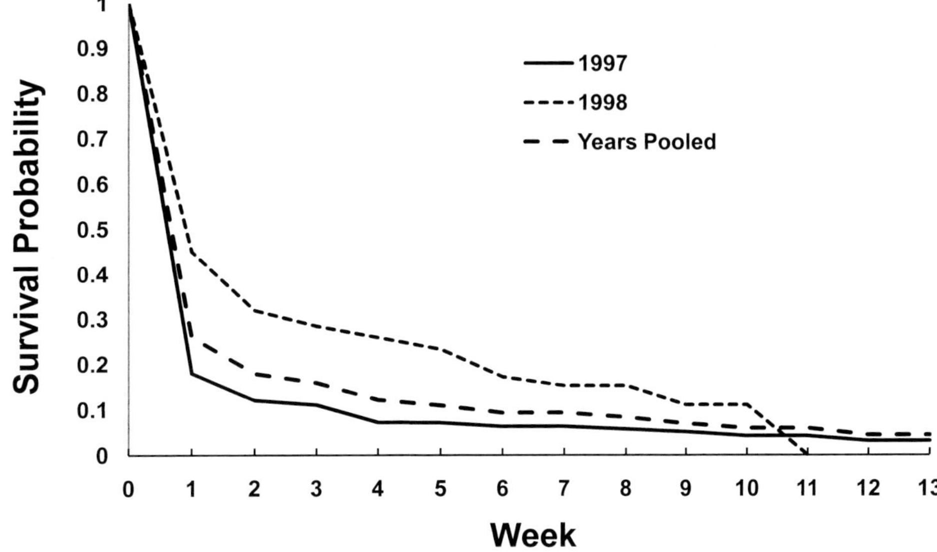

Figure 4.2

Survivorship of ruffed grouse chicks in the Appalachians, 1997–1998, using data sampled from intensively sampled broods (Haulton 1999).

Our observed survival of ruffed grouse chicks in the Appalachians appears to be much lower than that reported in previous studies. Survivorship to five weeks of age ranged from 0.06 to 0.19, and was 0.07 at 10 weeks post-hatch for data combined over 1997–1998. For comparison, chick survivorship to approximately 8–10 weeks post-hatch was 0.37 in New York (Bump et al. 1947), and was 0.51 for 12-week-old chicks in Alberta (Rusch and Keith 1971). One thing to consider is that the survival estimates in the New York and Alberta studies were based on flushes of hens with surviving broods and did not account for hens that lost their entire brood earlier in the summer, which creates a bias that inflates brood count averages. A recent study in northern Michigan did account for whole-brood loss by radio tagging grouse chicks and reported a chick survivorship of 0.32 at approximately 11–12 weeks post-hatch (Larson et al. 2001). Based on these results early in the ACGRP, it was obvious that chick survival was an important factor in grouse life history in the Appalachians.

The low survival in the first week of life is primarily a result of high incidence of whole-brood loss, with nearly one-third of broods hatched on VA2, VA3, WV2 and NC1 lost during the first week. Although reported to occur in other studies, whole-brood loss during the first couple of weeks appears to be more common in the Appalachian region. For broods that survive the first week, early mortality still appears to be higher in the Appalachians than elsewhere. By early July (five weeks of age), the average surviving grouse brood in the southern Appalachians consisted of three chicks. In the Great Lakes region, studies have reported 6–7 chicks per brood when grouse populations were high, and averaged 7–8 chicks per brood when populations were considered "low" (Dorney and Kabat 1960, Kubisiak 1978).

Telemetry Study on Broods. Because our early results indicated that the low survival of grouse chicks might be contributing to low population sizes, we undertook a more intensive study of brood survival using radio telemetry on chicks. Transmitter size and lack of reliable attachment methods have limited researchers' ability to examine survival and causes of mortality for ruffed grouse chicks in the past. Until recently, researchers have studied effects of factors such as weather, food availability, and predation on grouse chicks without the ability to track individual chicks through time. Recent advancements in transmitter technology such as miniaturization and improved attachment methods have allowed researchers to examine survival and causes of mortality in precocial chicks (i.e., chicks that are able to leave the nest and feed themselves immediately after hatching). During 1998–1999, Dobony (2000) initiated a pilot study to configure a transmitter design to study chick mortality from two to three days post-hatch to five weeks post-hatch (Figure 4.3). Smith (2006) then used the transmitters to examine chick survival and cause-specific mortality at WV1, VA2, and PA1 (Figure 4.4).

Using information from radio-tagged chicks we discovered a considerable amount about causes of mortality, survival rates, and overall ecology of grouse chicks in the Appalachians. Over three years, we captured 177 chicks from two to four days of age from 50 broods. We equipped 139 of these chicks with transmitter collars to monitor survival and identify causes of mortality. Ruffed grouse chicks selected to receive radio transmitters weighed on average 14.2 grams (139 chicks; range of weights 9.9–21.3 g) when captured. We determined the fates of 118 of 139 (85%) radio-collared chicks. A total of 110 (79%) chicks died and eight (6%) chicks survived to 35 days of age (Figure 4.5). Death from exposure and predation were the two main causes of known chick mortality across all study areas, and were likely underestimated because a number of individuals were censored from the analysis because we did not know their fate (the battery may have died or the transmitter may have been destroyed or taken a long distance by a predator).

We were able to document many predators of ruffed grouse chicks. Avian predators of young chicks included broad-winged hawks, red-shouldered hawks, and red-tailed hawks. Nests of these three species were located because grouse chicks with transmitters were fed to nestling hawks (Figure 4.6). On one occasion, three of four collared chicks were killed on the same morning by an unidentified mammal within a three-meter circle.

Figure 4.3

Left and center: Dobony (2000) initially tested two transmitter attachment methods for ruffed grouse chicks in the ACGRP. The transmitters are within the red circles, and the thin, black radio antennae can be seen in each photo. The glue-on method (left) did not prove as effective as the collar-type method (center), and was therefore abandoned in studies from 1999–2002. *Photos: Chris Dobony*

Right: We captured these two three-day-old ruffed grouse chicks as part of our survival study. Note the transmitter's antenna from the chick on the right and the U.S. quarter placed in the photograph for scale. *Photo: Brian W. Smith*

Figure 4.4

Top: Can you find the ruffed grouse chicks in this photo? As part of the ACGRP, we located broods and attached transmitters to individual chicks to determine survival rates and causes of mortality. To help you find them, the chicks are facing away from the camera and three black radio antennae can be seen if you study the photo closely. There are four chicks in the photo, and three have radio transmitters—challenging, isn't it? *Photo: Chris Dobony*

Bottom: Once broods were located and chicks captured, we randomly selected a portion of the chicks to receive a radio transmitter. In this photo, author Brian Smith is attaching a collar-type transmitter to a three-day-old grouse chick at WV1. *Photo: Rebecca Smith*

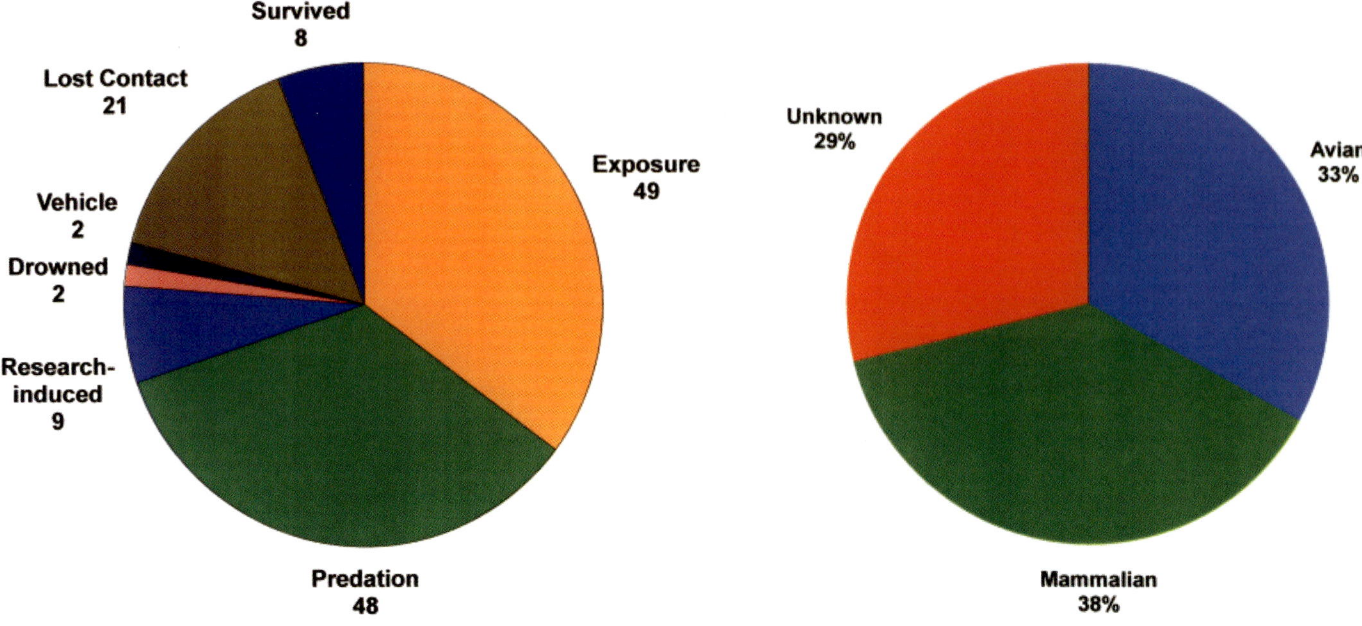

Figure 4.5

Left: Fate of 139 ruffed grouse chicks fitted with radio transmitters to determine causes of mortality and survival rates to five weeks of age at WV1, PA1, and VA2 from 2000–2002. Numbers indicate the number of times that particular fate occurred (e.g., eight chicks survived to 35 days of age over the three years).

Right: Predation of ruffed grouse chicks accounted for 44% of known mortalities, but rates among predation types were fairly even during the three-year period. Both graphs are based on data from Smith (2006).

Figure 4.6

Left: This radio transmitter (center of photo) belonged to an 11-day-old ruffed grouse chick from WV1 in 2001. The chick was scheduled to have its radio collar changed the same morning it went missing from the brood.

Right: The transmitter's signal led us to a red-shouldered hawk nest several hundred meters away from where the brood had been foraging for several days. The transmitter was actually in the hawks' nest, which had four nestlings near fledging age (only three appear in this photo). *Photos: Brian W. Smith*

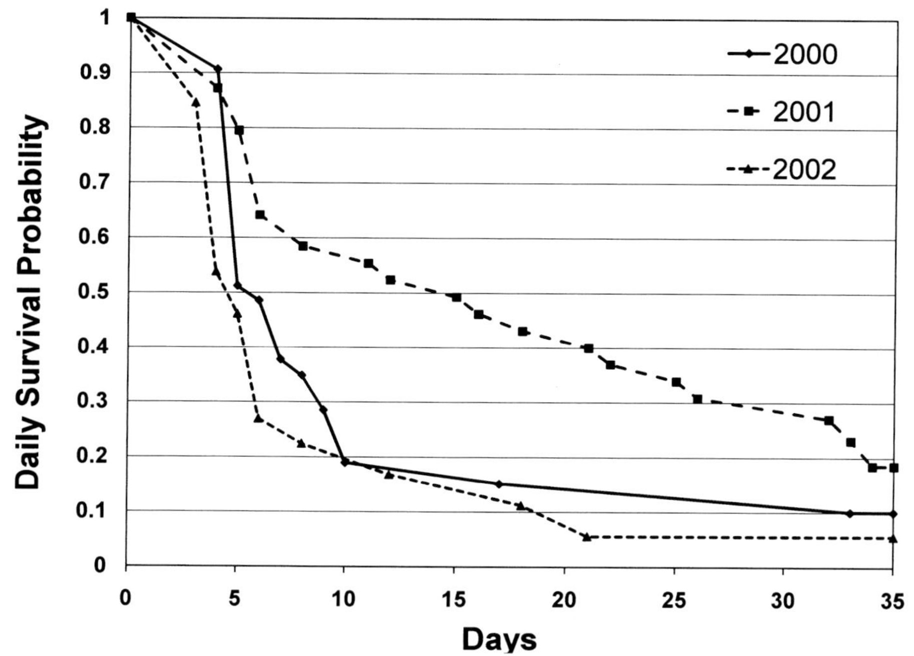

Figure 4.7

Survival estimates of ruffed grouse chicks for the first five weeks of life for three years. Survival in 2001 was clearly higher than the other two years, indicating that ruffed grouse chicks survived longer on average in 2001. Chicks were from WV1, PA1, and VA2 (Smith 2006).

Given the circumstances, the brood likely was attempting to remain concealed, but the predator located most of the chicks. We also recorded at least 14 (29%) entire brood losses out of 48 total broods during this study.

We calculated survival estimates using the information we gathered from the broods where we attached transmitters to individual chicks. We found no differences in survival among study sites over the entire study area (VA2, PA1, and WV1 all had similar survival rates, on average). However, we did find differences in survival between years within individual sites. For example, survival at PA1 in 2002 was much lower (0.0%) than survival on the same area in 2001 (23%), and survival at VA2 was 12% in 2000, whereas it was 0.0% in both 2001 and 2002 (Table 4.2). When considering survival rates by year, we found that chicks in 2001 survived longer on average and more frequently survived to 35 days than in 2000 and 2002 (Figure 4.7).

Survival rates of radio-collared chicks to 35 days post-hatch ranged from 0–0.23, with each site experiencing a survival rate of 0.0 during at least one season (Table 4.2). Survival estimates for the three years for ruffed grouse chicks were lowest at VA2 (0.09), and were similar at WV1 (0.12) and PA1 (0.13).

Table 4.2. Survival probability to five weeks of age for ruffed grouse chicks captured at 2–4 days of age and equipped with radio transmitters on WV1, PA1, and VA2 from 2000–2002. Chicks were monitored from time of capture until mortality or 35 days of age, whichever came first.

	Survival Probability		
Year	WV1	PA1	VA2
2000	0.00	0.10	0.12
2001	0.17	0.23	0.00
2002	0.12	0.00	0.00

So, what did we learn from this regional chick survival study that used radio telemetry? We estimated ruffed grouse chick survival during a critical time period—the first few weeks of a grouse's life. As noted, some methods to estimate chick survival (e.g., flush counts) can be biased for numerous reasons, and an accurate estimate of chick survival for the Appalachians was needed. Our method of radio tracking chicks was similar to a recent study in Michigan, but provided additional days of monitoring during the critical first week of life by placing collar-type transmitters on ruffed grouse chicks that were only two to four days old. Our observed survival rates (0.10, 0.19, and 0.06 for 2000, 2001, and 2002, respectively) in the Appalachians were much lower than those found in the Michigan study (0.29 in 1996, 0.32 in 1997). We also observed a very high incidence of entire brood loss (29%) and deaths caused by exposure (44%) compared to northern grouse populations. Overall, predation and exposure were the leading causes of known mortalities in the Appalachians (44% for each), and predation rate by avian predators (hawks and owls) and mammals were similar. In the Appalachians, ruffed grouse chick survival, and potentially grouse population size, may be limited by several factors including, but not limited to, the availability of high-quality brood habitat, effects of weather, diversity and abundance of avian and mammalian predators, and nutritional quality of their highly varied diet. Although our chick study provided valuable information, there is still much to learn about chick survival and its effects on grouse populations in the Appalachians.

Brood Movements and Habitat Use

Brood habitat has been considered the most specialized and most important cover type sought by grouse during their annual life cycle (Bump et al. 1947, Berner and Gysel 1969, Bergerud and Gratson 1988). Hens must lead newly hatched broods to areas that offer the right mix of ground cover to attract insects (their primary diet until around five weeks of age), but not so dense that chick movement is impeded. In addition to providing forage for chicks and brooding hens, brood habitat must provide cover from predators and inclement weather. In the northern grouse range, high-quality brood habitat has been described as containing ample herbaceous ground cover and high stem densities, usually associated with alder or aspen stands (Godfrey 1975b). Prior to the ACGRP, relatively little was known regarding brood habitat in the Appalachians, with descriptions ranging from regenerating stands (sapling and pole size) in upland and cove hardwoods to mature oak-hickory stands. It had also been suggested that the abundance and availability of brood habitat in parts of the Appalachians might be a factor limiting ruffed grouse populations. An overall lack of high-quality brood habitat would increase the chances of chick mortality and therefore negatively impact recruitment into fall populations.

Availability and distribution of high-quality brood habitat is important to ruffed grouse populations whether in the Great Lakes states or the Appalachians. High-quality brood habitat near the nest site may improve chick survival by reducing the need for the brood to travel far and wide searching for suitable cover. Although a newly hatched grouse chick can survive for a few days on yolk-sac reserves from the egg, their chances of survival would increase if they did not have to expose themselves to the risks of long-distance travel (e.g., predators, burning energy reserves, moving through unsuitable habitat) and could begin feeding immediately in suitable cover. Because we knew little about brood requirements throughout the Appalachians, several components of the ACGRP addressed brood movement and habitat requirements.

Movement from Nest Sites. We gathered habitat data at locations where we flushed grouse broods, and as a by-product of flushing broods frequently during the first five weeks of their life, we were able to document their movement patterns across the landscape. In some cases, we were able to determine how these movements might have influenced their likelihood of surviving or finding food.

During the intensive flush count study on VA2, VA3, and WV2, we followed 11 hens with broods during the first 12 days after hatching to track how far and how quickly they moved from nest sites (Haulton 1999). On average, hens with broods traveled 1,036 meters from their nests within the first 12 days, but this varied widely from hen to hen, ranging from only 262 meters to 2,145 meters. Hens with broods moved 146 meters from their nests per day on average, again with a wide range of estimates (21–178 m per day). At NC1, we found that average weekly distance from nest locations was not different between years, but the distance from the nest site increased over time in both years (Fettinger 2002, Figure 4.8). Unfortunately, no broods at NC1 survived past the third week in 2000, so we do not have a complete pattern for those two years. Figure 4.9 shows two examples of movement patterns hens with broods

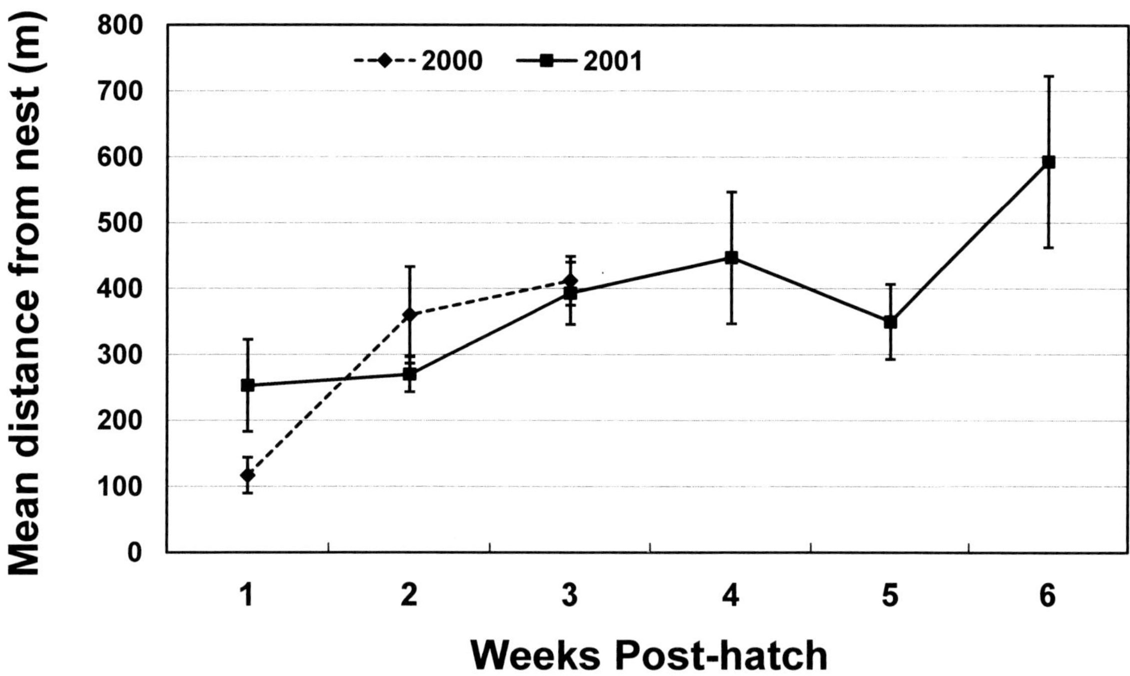

Figure 4.8

Average distance of ruffed grouse broods from their nest site in relation to time since hatching. These data were collected on NCI from 2000–2001 (Fettinger 2002).

undertook away from their nests in the mountains of North Carolina. The summary of data collected during 1997–2000 from seven ACGRP study areas found that broods moved from 41 meters to 689 meters away from their nests within one week after hatching. An interesting finding was that chick survival decreased as the distance traveled away from the nest increased during the first week (Tirpak 2000). Survival of chicks in broods that moved less than about 150 meters from their nest site during the first week was 60% or more, but survival of chicks in broods that moved more than 500 meters during the first week was less than 40%. After the first week post-hatch, we did not see a similar relationship of survival to distance moved from the nest site. Clearly, when conditions force a hen and her chicks to move long distances soon after hatching, high survival is not likely. Why grouse broods traveled so far shortly after hatching, what type of habitats were they looking for, and what they were dying from along the way became the emphasis of many of the remaining ACGRP projects.

Home Ranges. Movement distances from nest sites are closely related to brood home ranges, the areas that a hen and brood repeatedly travel across in their search for food and cover. Previous studies have shown that home ranges of females raising broods were larger than those of females whose reproductive attempts failed, and that home ranges of grouse in the Appalachians were larger than those in the Great Lakes. However, little information about home ranges during the brood-rearing phase existed for the Appalachian region.

Our analyses indicated that individual hens had relatively large home ranges in years they had successful broods (39 ha average), but in years they lost their brood, hens would restrict themselves to the smallest home ranges we observed for females at any time (15 ha average). In areas dominated by the oak-hickory forest association, as broods matured they made more use of hollows and bottomlands than the drier upland areas. As they made this habitat shift, home range sizes decreased. Bottomlands have been identi-

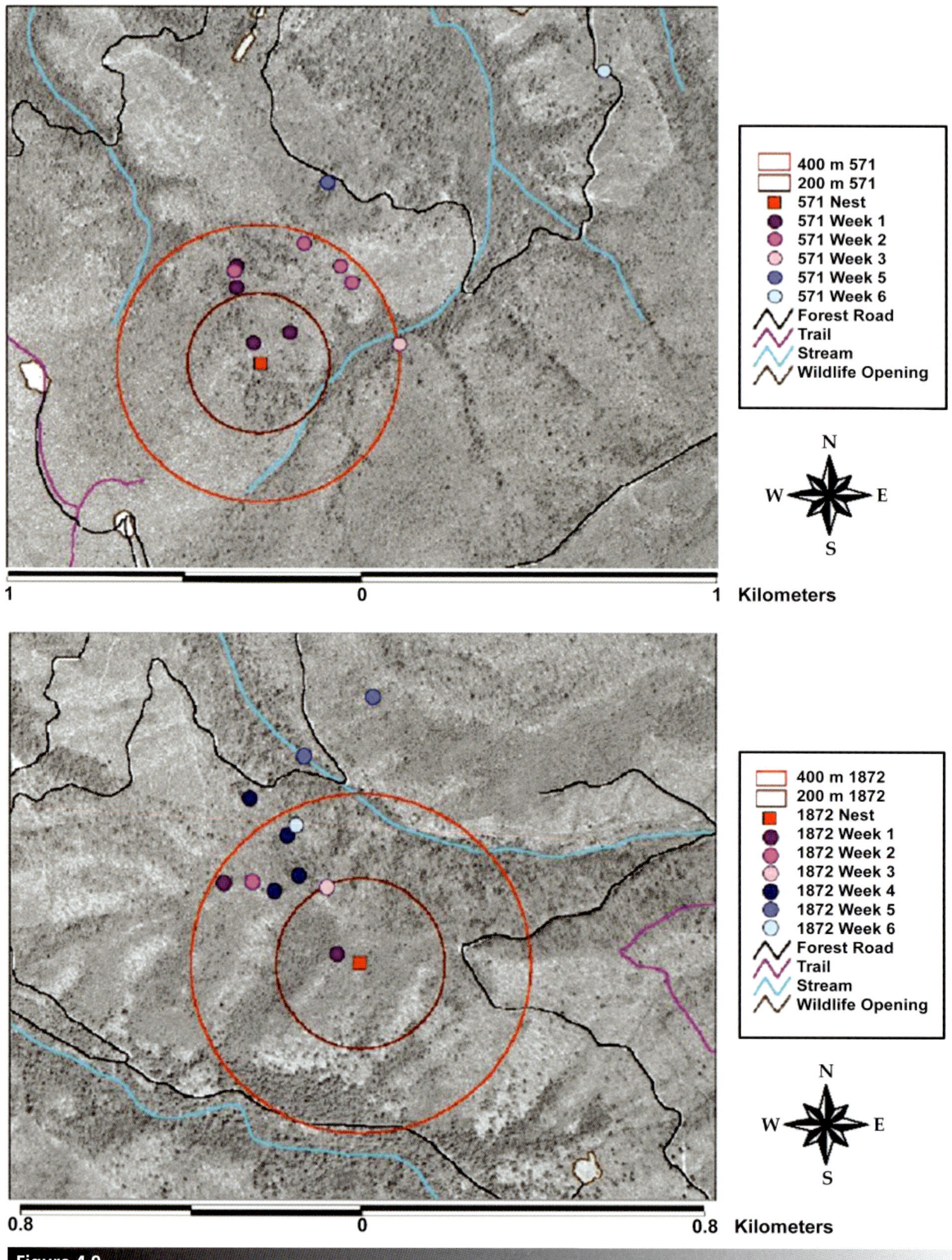

Figure 4.9

Examples of brood movements over the first few weeks of their lives, recorded on NC1 in 2001. If chicks survived, hens would typically extend the distance from their nest sites as broods aged (Fettinger 2002).

fied in previous studies, as well in the ACGRP, as preferred brood habitat in the southern portion of ruffed grouse range. The understory vegetation in hollows and bottomlands typically provides broods with better foraging and escape cover not found on dryer uplands in oak-hickory forests.

Cover Types and Habitat Structure. We had a number of questions related to brood movement and home ranges that likely were tied to habitat use. Previous studies throughout the range of ruffed grouse indicated the two most commonly reported characteristics of brood cover were high stem densities in regenerating stands and herbaceous ground cover that provides food and cover, but is not too thick to impede movement of young grouse chicks. The sparse information that existed for brood habitat in the southern Appalachians was for regenerating upland and cove hardwood stands in northern Georgia to mature oak-pignut hickory stands in North Carolina. We needed to describe brood habitat as thoroughly as possible, examining how habitat use changed as chicks aged. For instance, older broods primarily consume fruits and vegetation, and therefore may select cover types different from those chosen by younger broods that primarily consume arthropods.

Overall, percent ground cover and ground cover height were the most important characteristics of brood locations on the Virginia, West Virginia, and North Carolina sites. Selection for sites with higher percent ground cover and ground cover height was apparent throughout the first six weeks of the brood period but was less pronounced during the late brood period. Hens with broods typically selected sites with at least 60% ground vegetation cover that was at least 15 centimeters tall. Considering the size of the grouse hen and chicks, and their vulnerability as ground-dwelling birds to diverse predators, it makes sense that brood site selection be made on the basis of ground cover.

Abundance of vegetative foods was highest in brood plots during the early portion of the brood period. This was not expected since a chick's diet is still primarily arthropods at this time. We may have found this because a hen's diet is dominated by fruit and forage, and she may be benefiting from an abundance of forage plants and that most ground cover plants also provide forage and fruit. Most studies outside the Appalachians report brood sites with high numbers of woody stems. On VA2, VA3, and WV2, brood sites were best described as having slightly fewer trees (about 430 trees/ha) than generally present in the forest (490 trees/ha). However, midstory stem density was greater on brood sites compared to what was available on average in North Carolina.

One component of the ACGRP examined small-scale habitat features that influenced brood survival at seven sites (PA1, KY1, MD1, VA2, VA3, WV2, and WV3) from 1997–2000. We recorded habitat data at sites located during brood flushes at one, three, and five weeks of age. We measured the amount of tree cover (all deciduous and coniferous trees greater than 7.6 cm dbh), groundcover, dead woody material, stem density, woody stems less than 7.6 cm dbh, and basal area. Interestingly, this analysis showed no relationship of habitat to three different survival classes (low: 0–0.379; medium: 0.380–0.670; high: 0.671–1) during the first week or fifth week of broods' lives. However, percent ground cover was lower in areas used by three-week-old chicks in the high survival group when compared to the low survival group at the same age. Average ground cover ranged from 57–87%, and no broods were found at sites with less than 35% ground cover. This study confirms the importance of ground cover dense and diverse enough to attract insects for forage and provide cover for broods, but not so dense that it impeded travel and feeding behaviors of broods. In North Carolina, dense understory conditions created by perennial cool-season grasses such as orchardgrass prevented chick movement through these managed herbaceous openings; however, broods were observed foraging along their periphery. Herbaceous and woody stem cover provided by various forbs, brambles, shrubs, and regenerating hardwoods created desirable conditions for foraging and concealment along margins of managed openings (Jones 2005).

Based on our telemetry data, we found the selection of bottomlands by females was stronger in oak-hickory forests when compared to mixed-mesophytic forests, and was strongly associated with selection for access routes such as trails and paved, unpaved, and vegetated roads. Bottomlands and access routes have been considered to be important foraging sites for broods, and our telemetry data showed this pattern was evident in oak-hickory forests during the breeding season. Oak-hickory forests typically occur on much drier, poorer soils than mixed-mesophytic forests, so it is likely that grouse broods are using bottomlands on oak-hickory sites because they are moist areas that support lush vegetation (unlike the uplands), much like mixed-mesophytic forests.

So, what habitat conditions do ruffed grouse broods in the Appalachian Mountains need to thrive? The primary structure needed by grouse broods is a well-developed and diverse ground cover that provides overhead cover, but open enough underneath to permit travel. Use of early successional and young forest stands is important in many cases because of desirable cover and food availability. Invertebrates are more available to chicks on sites with abundant ground cover vegetation and soft mast is usually more abundant in these areas. In oak-hickory forests, bottomlands and riparian zones usually supply these conditions selected by broods.

Invertebrate Abundance. We have discussed only briefly what grouse chicks eat in relation to the habitats they use, but their diet and its availability is an important component of brood habitat and chick survival. Many researchers have emphasized the importance of available, high-quality foods for ruffed grouse broods and have found that chicks require a diet with approximately 30% protein to accommodate rapid tissue growth during the first several weeks after hatch. If a high-quality diet is available, developing chicks can double their weight within the first week and increase their chances of survival. Insects and other invertebrates are high-protein items that are hopefully abundant and available to grouse chicks. These invertebrates on the ground cover and in the leaf litter may make up over 90% of ruffed grouse chicks' diet during their first three weeks after hatching. After that, fruits and green leaves become more common in their diet and by the eighth week, chicks are feeding more on vegetation and fruits than invertebrates. Habitats that provide a high number and diversity of "bugs" early in the brood period may not offer the type or abundance of plant foods they'll need a few weeks later. Broods typically show a change in habitat characteristics from the early brood period to the later brood period (likely a result of a change in dietary needs), but we had little information about food abundance and diversity as it related to habitat characteristics. Therefore, we also investigated invertebrate communities present in brood habitats.

Our first invertebrate study was conducted in conjunction with the brood habitat study at VA2, VA3, and WV2 during 1997–1998 (Haulton 1999). Using a variety of methods, invertebrates were collected at known brood locations and nearby random plots by sampling the top of the leaf litter, catching low-flying insects, or "sweep-netting" those clinging to ground cover. Grouse chicks rarely "scratch" or turn over leaves to find insects (they are mainly attracted by movement), so sampling was limited to what a grouse chick might encounter with little effort. We did not find a difference in invertebrate abundance in litter samples when comparing brood and random plots. However, through our "sweep-netting" samples, we found invertebrates flying or clinging to low vegetation were more abundant (about 56 per sample) in plots used by broods than in random plots (10 per sample). Additionally, we found that brood plots supported on average more invertebrates than random plots for all weeks during the first four weeks of a brood's life. Abundance of individuals in the arthropod Orders Coleoptera (beetles), Homoptera (leafhoppers), and Arachnida (spiders, ticks) were higher in brood plots during the first three weeks of the brood season. In North Carolina we found the density of preferred orders, primarily ants (Hymenoptera) and leafhoppers (Homoptera), was greater on brood plots compared to random locations. Tying this back to the brood habitat sampled during this study, the open midstory and well-developed ground cover that was observed at brood sites in the first six weeks not only provided cover for hens and their chicks, but provided abundant invertebrates as well.

The second study of invertebrate abundance

and diversity took place during 1998 on WV1 (Dobony 2000). We investigated how temperature, rainfall, and cover type influenced invertebrate abundance, biomass (essentially, the total dried weight of all invertebrates captured by cover type), and the total number of arthropod families collected in each cover type. We found that roads lacking close forest cover produced higher numbers of arthropods than any other cover type except regenerating mesic-Allegheny hardwood stands that were 6–15 years old. Biomass of invertebrates was similar in all cover types, except that Allegheny hardwood regenerating stands that were 6–15 years old produced more invertebrate biomass than open-cutovers that were less than two years old. In terms of environmental conditions, we found that as temperature increased, we captured more invertebrate biomass, but abundance did not increase. Also, rainfall had a negative impact on both invertebrate abundance and biomass; as rainfall amounts increased, both the number and biomass of invertebrates we captured decreased. Based on these findings, it appears that managing non-forested roads and creating openings would provide both the ground cover structure and diversity needed to support invertebrate communities and provide cover to ruffed grouse chicks as they foraged.

The final study of invertebrates as they related to brood habitats took place at NC1 from 2000–2004 (Jones et al. 2008). Arthropods were sampled at brood locations and nearby random plots during the first six weeks after hatching. At NC1, we used a "terrestrial vacuum sampler" (Figure 4.10), which is a device that pulled arthropods from the air, vegetation, and top of the leaf litter very efficiently. Total invertebrate density was greater at brood locations than random plots. The late brood period did not differ in overall invertebrate density, but the density of preferred invertebrate orders as identified in other studies was greater at brood locations than at random locations. Preferred orders had higher densities at brood locations during both the early and late periods.

Our work on broods clearly showed that the period from hatching to the end of the summer is a critical time for ruffed grouse chicks. Mortality was high, especially in the first week after hatch, and this has the potential of being a limiting factor to populations. We found some clear habitat preferences during the brood rearing period, with hollows and bottomland areas that have a relatively high percentage of herbaceous ground cover being preferred. These sites also tended to support more insects and other invertebrates, which are important for broods as food during their critical first weeks of life. Habitat management efforts that would emphasize these characteristics presumably will provide conditions that will contribute to survival of hens and their broods during this critical period.

Figure 4.10

A Terrestrial Vacuum Sampler was used at NC1 to sample insects and other potential food items of ruffed grouse chicks. *Photo: Craig A. Harper*

5 Survival

Patrick K. Devers,
Gary. W. Norman,
Dean. F. Stauffer,
and David. E. Steffen

For a ruffed grouse, "just trying to survive" is not a matter of simply commuting to the office, writing some emails while the boss isn't looking, and whipping up a quick dinner before retiring to a warm comfortable bed. Rather, it is the difficult task of negotiating the landscape to find a few acorns or grapes here and there and finding a dense stand of shrubs or trees to roost in and get out of the cold blowing rain and sleet, all while trying to evade skilled predators, such as the Cooper's hawk, intent on securing their own supper. For ruffed grouse, just trying to survive is a daily challenge.

Survival is one of four vital characteristics of a population (the others being reproduction, immigration, and emigration) that influences wildlife population dynamics. The relative importance or influence of survival on population dynamics varies among species and even populations. Those who study population dynamics generally classify species into one of two life-history strategies. The first strategy, referred to as *K-selection* (or *K-selected*), is characterized by species that have high annual survival and low reproductive rates. Typically, the population growth rate of K-selected species is influenced more strongly by changes in juvenile and adult survival than in reproduction. The other strategy, referred to as *r-selection* (or *r-selected*), is characterized by species that typically have low annual survival and have high reproductive rates. In between the two idealized extremes lay most species, and it is important to evaluate the life-history strategy of a species or population on a continuum between r- and K-selection.

Grouse species, including the ruffed grouse, are typically considered to lie towards the r-selected end of the life-history continuum. However, a review of ruffed grouse ecology suggests grouse species exhibit two survival modes throughout their distribution. One is characterized by populations with annual survival of over 55%, and the other by annual survival less than 55%. For example, ruffed grouse annual survival in the northern portions of its distribution (e.g., New York, Wisconsin, and Alberta) ranged from 25–50% (Bump et al. 1947, Rusch and Keith 1971, Small et al. 1991). In contrast, in the more southerly portion of ruffed grouse range, annual survival has been reported to be 47–62% (Triquet 1989, Swanson et al. 2003). These findings suggest ruffed grouse populations may experience different influences (i.e., differences in habitat quality, types and abundance of predators, weather conditions, and human influences) and these differences in outside factors result in slight shifts of life-history strategies.

Humans can influence ruffed grouse populations through a variety of activities. A hunter and his dog working though young stands of regenerating hardwoods during the short, crisp winter days presents perhaps the most notable form of human activity that influence ruffed grouse survival. Hunting of wildlife populations is founded on the *Compensatory Mortality Hypothesis* first proposed by Paul Errington in 1935 (Errington and Hamerstrom 1935). Based on his work with northern bobwhite and muskrat, Errington hypothesized that a piece of land could provide enough shelter, cover from predators, and food for a limited number of individuals during the most severe season of the year (typically winter in North America). This characteristic of the land is referred to as the *carrying capacity*. However, wildlife populations tend to exceed the carrying capacity during spring, summer, and fall due to production of young, thus producing a *doomed surplus* that will perish during the winter period from a variety of natural sources including predation, starvation, and disease. Under these conditions, the compensatory mortality hypothesis states human harvest of wildlife populations replaces one or more natural mortality factors (e.g., predation) without reducing the breeding population. This concept forms the foundation for regulated sport harvest in North America and has been supported by multiple field studies that have indicated a harvest rate of 25–50% of the preseason ruffed grouse population is compensatory (Bump et al. 1947, Monschein 1974).

For every grouse that goes into the hunter's bag, several escape with an explosive and heart-stopping flush and smooth glide through the forest, reminding the hunter why the ruffed grouse is considered the "King of Gamebirds." But are the bagged grouse really just the unfortunate members of the doomed surplus that were destined to die before winter's end? Or, is it possible they would have survived to drum on a downed, moss-covered log in the early morning spring hours or incubate a clutch of cream-colored eggs until the 8–12 chicks hatch in near perfect synchronicity 24 days later? If these bagged grouse would have otherwise survived the winter to breed in the spring, wouldn't harvest cause the population to decline? This scenario is referred to as the *Additive Mortality Hypothesis,* which predicts each death (natural or harvest) adds to total mortality and reduces population size. Several researchers have concluded that harvest rates of over 30% of the preseason population is additive and may cause a decline in ruffed grouse abundance (Fischer and Keith 1974, Small et al. 1991).

Prior to the initiation of the ACGRP, few researchers had investigated ruffed grouse survival and the effects of hunting in the Appalachians. Ruffed grouse management, and particularly harvest management, was based on research conducted in the northern portion of ruffed grouse range where populations are high and subject to a 10-year cycle. But management prescriptions from the core of ruffed grouse range may not be applicable in the Appalachians due to differences in weather conditions, food resources, predator communities, and harvest management. A goal of the ACGRP was to investigate ruffed grouse ecology, including survival and the effects of hunting in the Appalachians and provide regional recommendations for ruffed grouse management. In the following pages we report on several aspects of ruffed grouse survival, including estimates of annual and seasonal survival rates, causes of mortality, and the effect of hunting.

Methods

In addition to the general description of our study areas provided in Chapter 2, we feel it is important to review here the differences between oak-hickory (OH) and mixed-mesophytic (MM) forest associations (Figures 2.2 and 2.3). The primary difference between these forest associations, at least as it relates to ruffed grouse ecology, is the abundance and composition of tree species, including members of the white and red oak groups, a variety of hickory species, birch, cherry, and aspen. Oak-hickory study areas, as the name implies, were dominated by oaks and hickories, but perhaps more importantly, birch, cherry, and particularly aspen, were rare or absent. In contrast, MM study areas supported an abundance of oaks and hickories, but also include beech, birch, cherry, and aspen. The buds of birch, cherry, and aspen are a relatively high-quality food resource for ruffed grouse, particularly during late winter. Importantly, buds of these tree species are easily accessible and available every year. Acorns of white and red oaks are also a high-quality food resource, but the availability of acorns is highly variable across the landscape and from year to year. The difference in tree composition and abundance in oak-hickory and mixed-mesophytic forests suggest important differences in the quality and availability of food resources for ruffed grouse.

Data on survival and cause-specific mortality were collected using radio-telemetry, as described in Chapter 2. We monitored grouse at least two times per week to determine their status as alive or dead. When a mortality signal (i.e., doubling of the transmitter pulse rate) was detected, field personnel would use the radio-telemetry equipment to locate the radio collar and grouse remains. Once the radio collar was recovered, the field biologists would search the immediate area for field sign to determine the cause of mortality as avian predation, mammal predation, un-identified predator, or other. Based on the last day the bird was known to be alive and the day the carcass was recovered it was possible to estimate the day the bird died. Information on hunter-killed birds was obtained using a $25 reward inscribed on the transmitter. Any hunter who harvested one of our birds would receive $25 for simply calling the number on the transmitter and reporting the day and location where the bird was harvested.

The ACGRP was an uncommon project in wildlife research because of its experimental design, geographic scale, and duration. Many researchers have previously investigated the effect of hunting on ruffed grouse, and though these studies have provided a wealth of information and insight into ruffed grouse survival and harvest, the conclusions often have limited application because the studies lacked experimental control, replication, and manipulation of harvest. The ACGRP was designed to experimentally test the compensatory mortality hypothesis. To accomplish this, we evaluated ruffed grouse survival on seven study areas (KY1, MD1, WV1, WV2, VA1, VA2, and VA3—Figure 2.1) with and without hunting. During Phase I of the ACGRP (fall 1996 to summer 1999) each of these study areas were open to hunting following normal state hunting regulations. Hunting seasons typically ran from early October to late February with daily bag limits ranging from one to four grouse and possession limits of four to eight birds. During Phase II of the ACGRP (fall 1999 to summer 2002), we closed three study areas (KY1, WV2, and VA3) to hunting while allowing hunting to continue on the remaining four areas (MD1, WV1, VA1, and VA2). Those areas closed to hunting are referred to as the "treatment group" and the areas that remained open to hunting are referred to as the "control group."

Using data collected via the use of radio telemetry, we estimated annual and seasonal survival rates and the influence of multiple factors on ruffed grouse survival. Details concerning our methods are described further in Devers (2005).

Results and Discussion

Any experienced birder or grouse hunter can recount years when it seemed like ruffed grouse had all but disappeared, and yet there are other years when grouse seemed to be behind every other tree. And what can we make of reports of some hunters flushing several grouse in a single day in one hollow while their hunting buddies flush only one or two in another hollow? This apparent change in the abundance of ruffed grouse over time and space is in part due to changes in ruffed grouse survival. During the course of the ACGRP, from 1996 to 2002, ruffed grouse survival varied from year to year, within years from winter to summer, and across study areas. During the course of our study ruffed grouse annual survival ranged from 44% to 53%. Survival also varied with the changing seasons. As one would expect, survival was highest in summer (93%) when cover is abundant, the days are warm, and food is plentiful. In the fall, food was still plentiful and the days warm, but young grouse moved across the landscape searching for a place to settle in for the winter during a time known as the "fall shuffle." Adult grouse, particularly males, defend their territories from wandering juveniles, while raptors pass through the Appalachian forests in greater numbers on their southern migration at this time. During this period ruffed grouse survival decreased from the summer peak and ranged from 74% to 83%. Survival was lowest in winter, ranging from 72% to 84%, as grouse strived to find food—including shriveled grapes, leaves of evergreen species like greenbrier, or an overlooked acorn—or burrowed into a pile of fallen oak leaves to shelter themselves from the cold rains and winds. Survival increased in the spring from the winter low as the days grew longer and warmer, and early green growth of herbaceous species provided an important food resource and cover. During spring, survival ranged from 75% to 92%.

Our estimates of ruffed grouse annual survival and the observed trends in seasonal rates were similar, but slightly higher to previously reported rates from the northern core of ruffed grouse range (i.e., the Great Lakes and southern Canada region). In their classic study of ruffed grouse, Gardner Bump and his co-workers (1947) estimated mean survival was 42% and 50% on two study areas. In Wisconsin, annual survival ranged from 25% to 34% (Small et al. 1991). Further north in Alberta, Canada annual survival ranged from 27% to 30% (Rusch and Keith 1971). Notably, survival in more southerly areas, including Ohio (47%) and Kentucky (62%), were similar to our estimates (Swanson et al. 2003, Triquet 1989).

As would be expected from our estimates of annual survival, estimates of seasonal survival rates were also higher compared to estimates for the northern core of ruffed grouse range. However, the pattern of survival being highest in the summer and lowest in the winter appears to be the norm throughout ruffed grouse range. In central Wisconsin, adult and juvenile summer survival was 85% and 65%, spring survival was 73% and 50%, fall survival was 65% and 48%, and winter survival was 57% and 55%, respectively. In Alberta, winter survival was 42% and 67% in 1967 and 1968, respectively.

Day in and day out, ACGRP field biologists would listen to the slow and steady beep-beep of radio waves emanating from the transmitters hanging loosely around the necks of unsuspecting grouse. Switching between frequencies, biologists would systematically verify each bird was alive, confirmed by the same monotone *beep-beep*. Days and weeks could be long and tedious, particularly in the summer months when adult males moved short distances and could be heard in almost the same location every day.

But occasionally, we would be startled when we entered in a frequency expecting the typical slow beep-beep, only to hear a much faster pulse. If a bird failed to move in an eight-hour period, the pulse of the transmitter signal would approximately double, usually indicating the bird had died. Each time we heard this rapid pulse our heart rate would quicken as if it was trying to keep pace, and we would wonder, "What happened to this bird? Is it really dead, and if so, what killed it?" We then gathered our equipment and data forms and would head off into the woods moving over downed logs and crawling through thick patches of rhododendron, following the steady but rapid beeps until the signal strength was so strong we knew the collar was close. At this point the detective work began. We would search the immediate vicinity for signs of the bird or the transmitter and then slowly walk in ever-wider circles looking for any sign of the bird until we came across the transmitter. Often the condition of the transmitter and the bird's remains would provide clues as to the ultimate fate of the bird, such as a coiled antenna, which is a telltale sign of predation by an avian predator. While feeding on the grouse, raptors often strip the radio-antenna in their beaks causing it to coil like Christmas ribbon. Searching the immediate vicinity and the remains of the carcass would provide additional clues to allow us to determine the exact cause of death.

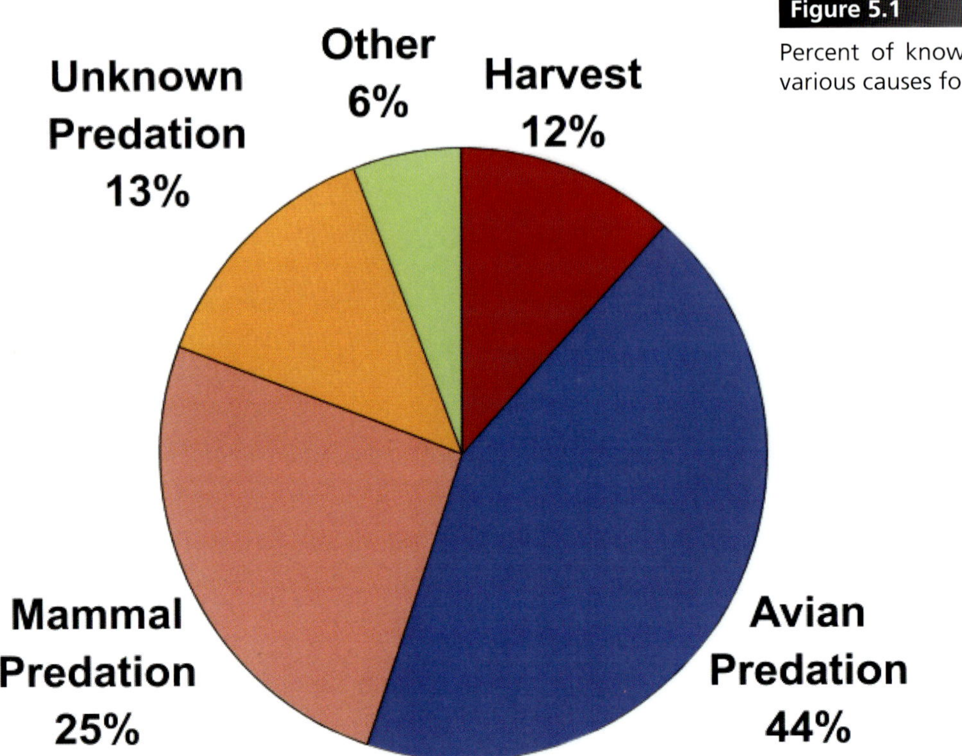

Figure 5.1

Percent of known ruffed grouse mortality due to various causes for the study period, 1996–2002.

Ruffed grouse die from a variety of factors including disease, accidents, predation, and hunting. During the course of the ACGRP, predation accounted for 84% of all known mortalities (Figure 5.1). Avian predators were the leading cause of predation, followed by mammalian predators, and unidentified predators. Avian predation rates were highest in the fall and spring, coinciding with peaks in the abundance of migrating hawks (Bumann and Stauffer 2004, Chapter 6). Causes of ruffed grouse mortality in the Appalachian region were similar to those reported throughout the range of ruffed grouse (Bump et. al. 1947, Marshall and Gullion 1965, Rusch and Keith 1971, Rusch et al. 1978, Small et al. 1991, Swanson et al. 2003). In the core of ruffed grouse range, northern goshawks and great horned owls are considered the primary predators of ruffed grouse, but goshawks are rare in the Appalachian region; the primary predators in the Appalachians are the Cooper's hawk and owls (Bumann and Stauffer 2004).

Legal hunter harvest accounted for only 12% of all known mortalities, a smaller portion compared to previous studies. Harvest accounted for 13– 20% of known mortalities in New York (Bump et al. 1947), 28% (Small et al. 1991) to 40% (DeStefano and Rusch 1986) in Wisconsin, and 19–48% in Alberta (Fischer and Keith 1974). Swanson et al. (2003) concluded harvest (8.6% of known mortalities) was a minor source of grouse mortality in Ohio.

So, what causes survival to change among areas and through time? Is it the abundance of predators, cold temperatures, age or sex of the bird, or food abundance? Is it some combination of these or other factors? Or, are changes in ruffed grouse survival tied to the 10-year abundance cycle that is so well documented in the Great Lakes region? Importantly, we did not find any evidence, nor have other researchers, that ruffed grouse in the southern and central Appalachians follow the 10-year cycle common to grouse populations in the Great Lakes region, but we were able to identify several factors that appear to influence ruffed grouse survival in the Appalachians. These factors include forest association, sex, and age.

Table 5.1. Comparison of ruffed grouse average population vital rates in oak-hickory and mixed-mesophytic forests in the southern and central Appalachian region 1996–2002.

Vital Rate	Forest Association: Oak-hickory Mean	Forest Association: Mixed-mesophytic Mean
Nesting Rate	86%	100%
Re-nesting Rate	3.2%	45%
Clutch Size	9.4 eggs	10.37 eggs
Nest Success	63%	70%
Chick Survival	21%	39%
Adult Annual Survival	≈50%	≈42%

First, and perhaps the most interesting, was the influence of forest association on ruffed grouse survival and other vital population rates (Table 5.1). Our data suggest survival was 3% higher (in all seasons) in OH forests compared to MM forests (Figure 5.2), and just for adults, survival was about 8% higher in OH forests (Table 5.1). We believe our results support the idea that grouse exhibit two survival modes (high and low) and that ruffed grouse in MM forests exhibit the low survival mode and those in OH forests exhibit the high survival mode. We believe the difference in survival in OH and MM forests is the result of ruffed grouse adapting under different environmental conditions.

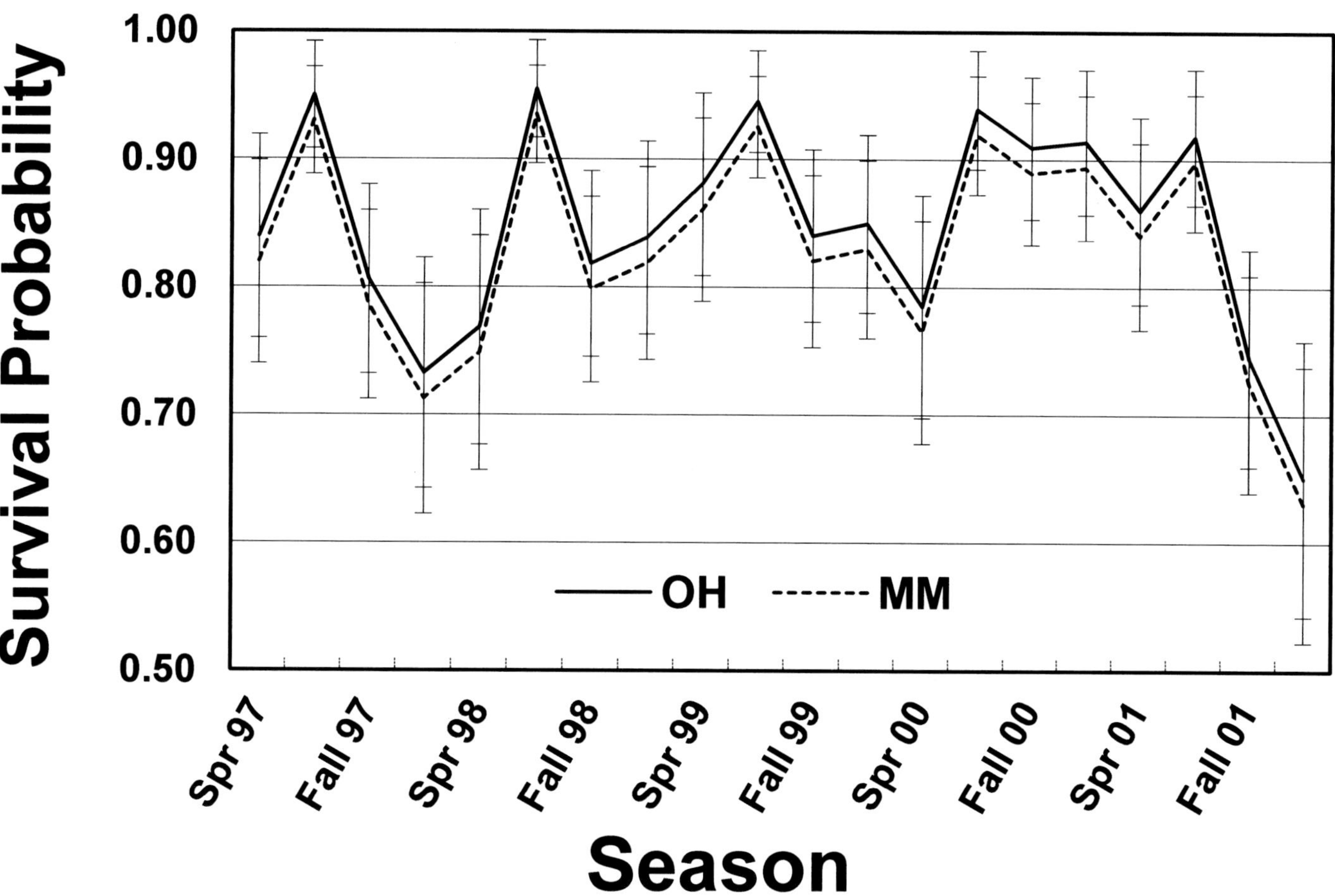

Figure 5.2

Ruffed grouse survival patterns in oak-hickory (OH) and mixed mesophytic (MM) forests in the Appalachian region for the period of 1997–2001.

Ruffed grouse in OH forests appear to experience periods of nutritional stress due to failed acorn crops in some years and the associated increased energetic demands. Under these conditions, ruffed grouse can best adapt by maximizing survival over reproduction, which demands substantial energy. In years of poor acorn production, ruffed grouse in OH forests may forgo nesting in order to increase the probability of survival and the potential to reproduce in the future. In contrast, birds inhabiting MM forests typically do not experience periods of nutritional stress because of the presence of high quality and consistently available food resources, including the buds of cherry, birch, and aspen trees. Under these conditions grouse can best adapt by maximizing their reproductive effort and success. This is evidenced in the fact that in MM forests, the nesting rate, renesting rate, clutch size, next success, and chick survival were all higher than in OH forests (Table 5.1).

Our data did not provide a clear indication of the relation between survival and grouse age. In most years and seasons grouse age did not appear to influence survival. However, in some years adults did have slightly (1–2%) higher survival in fall and winter than juveniles (3–15 months old). Previous research paints a similar picture. For example, adult ruffed grouse in Wisconsin (Small et al. 1991) and Alberta (Rusch and Keith 1971) had higher survival than juvenile ruffed grouse, but the opposite appeared to be true in Michigan (Clark 2000). Others working in Minnesota (Gutierrez et al. 2003), Ohio (Swanson et al. 2003), and Kentucky (Triquet 1989) concluded survival did not differ between adults and juveniles. We suggest age has only a minor influence on ruffed grouse survival, and probably operates during very short windows, specifically early fall (e.g., Sept.). Most importantly, the influence of age on survival varies across time and space. We suggest that variation in weather conditions, predator composition and abundance, and the condition of individual birds more strongly influences juvenile than adult survival and will cause juvenile survival to differ from adult survival in some years and some locations.

As for age, survival did not appear to differ between males and females except during spring in one or two years of the study. Other researchers working throughout ruffed grouse range concluded ruffed grouse survival does not differ between males and females (Rusch and Keith 1971, Gutierrez et al. 2003, Swanson et al. 2003). However, Bergerud (1988) proposed that males and females have different mortality rates because of differences in cost of reproduction. Reproductive effort is "cheap" for males, and is limited to advertising for females and mating. Thus, for males reproduction takes relatively little energy. On the other hand, females have to expend much more energy in reproduction, including producing eggs, incubation, defending nests, and caring for young. It is believed these added energy costs should result in lower female survival rates during the reproductive season. Others have argued that displaying males are at greater risk of predation from aerial predators and may have lower survival during the spring than females. However, for Appalachian grouse we conclude that during the course of the year, the influence of age and sex on survival is probably minor.

During the past five or six decades biologists and hunters have debated the effects of hunting on ruffed grouse and have strived to determine whether hunting causes ruffed grouse numbers to decline. Unfortunately, the answer to this question has been as elusive as any grouse in the autumn woods. Several studies support the additive mortality hypothesis, indicating hunting can cause populations to decline. Others have supported the compensatory mortality hypothesis suggesting that we can harvest less than 50% of the fall population without causing a decline in the number of breeding birds the following spring. Still others argue that harvest is compensatory up to some threshold level, after which point continued harvest would cause the population to decline. The ACGRP was designed to answer this question and provide harvest management prescriptions for the region by experimentally testing the effect of hunting of ruffed grouse survival.

Harvest rates experienced during our study were lower than reported in other parts of ruffed grouse range. The average harvest rate on control areas was 8% (range 4–13%) compared to 20% on treatment areas prior to closure (during 1997–1998 and 1998–1999). In comparison, mean ruffed grouse harvest rates in New York ranged from 13% to 20% (Bump et al. 1947) and 29% to 50% in Wisconsin (DeStefano and Rusch 1986, Small et al. 1991). Our results indicated that annual survival did not increase on the treatment areas after hunting was closed (Figure 5.3). This lack of a response supports that harvest rates of 8–20% are compensatory and did not cause a decline in the breeding population. We conclude that a harvest rate of less than 20% in the southern and central Appalachian region is compensatory. We believe current harvest rates can be maintained, but regional state agencies should not amend hunting seasons to facilitate higher harvest rates, particularly in light of the loss of habitat that is occurring throughout the region. It is important to recognize that we cannot assume harvest rates higher than those observed in this study are compensatory; nor can we extrapolate our results beyond the Appalachian region.

There is yet no conclusive answer as to whether hunting mortality is additive or compensatory in ruffed grouse. Kubisiak (1984) found that a mean harvest rate of 44% (range 23–72%) was additive to natural mortality and reduced ruffed grouse densities in Wisconsin. Yet, other studies suggest harvest mortality is compensatory up to a threshold and then becomes additive. In New York, researchers experimentally harvested 19.5%, 20%, and 13.4% of the fall population on one study area and compared over-winter survival to an adjacent reference area in three consecutive years. At the end of the study the researchers concluded that 50% of the preseason population could be compensated by a decrease in natural mortality and harvest mortality is a minor component in ruffed grouse population dynamics (Bump et al. 1947). In Wisconsin, ruffed grouse captured within 201 m of an access trail experienced higher harvest rates (48%) and lower annual survival (23%) than birds captured further away from the road (19% and 36% respectively) suggesting harvest mortality added to overall mortality for birds captured closer to access trails (Fischer and Keith 1974). However, independent estimates indicated the grouse population remained stable from fall to spring (October–May) suggesting harvest may have been completely compensatory (Fischer and Keith 1974).

Numerous studies have concluded harvest mortality is compensatory. In western North Carolina, grouse abundance did not differ before, during, or after hunting season in small woodlots with three levels of prescribed hunting pressure (no hunting, moderate hunting, and unrestricted hunting, Monschein 1974). Gullion and Marshal (1968) found hunters could harvest 18% of territorial male ruffed grouse in Minnesota without reducing the population. Others in Wisconsin suggested harvest mortality less than 40% of preseason population is compensatory (Dorney and Kabat 1960). In Ohio, harvest accounted for 8.6% of mortalities and was determined to be compensatory (Swanson et al. 2003).

A common factor in studies concluding either partial or complete compensation of harvest mortality was the role of immigration. Many studies have compared demographic rates and densities on hunted and non-hunted sites, but the results are not always conclusive because the researchers could not discount the possibility that birds moved into the study area from other populations, thus creating the appearance that hunting did not cause decline in the grouse population. Gullion (1983) argued that inaccessible areas (or limited access areas) could serve as refugia for ruffed grouse and supply surplus birds to areas that experience high hunting pressure. We did not have information on immigration of grouse into our study sites, and cannot discount the possibility that this could have contributed to the conclusion that hunting was compensatory on our areas.

Despite the extraordinary amount of effort exerted in terms of person hours, equipment, and funding during the ACGRP, we did face several limitations in the execution of the test of the compensatory mortality hypothesis. First, we had a relatively small sample (four control and three treatment areas). We did not use data from the other study areas because complete data were not collected during Phase I (1996–1999) and Phase II (1999–2002) of the study. Considering the inherent variability among study areas and years, and our small sample size, it is possible that we failed

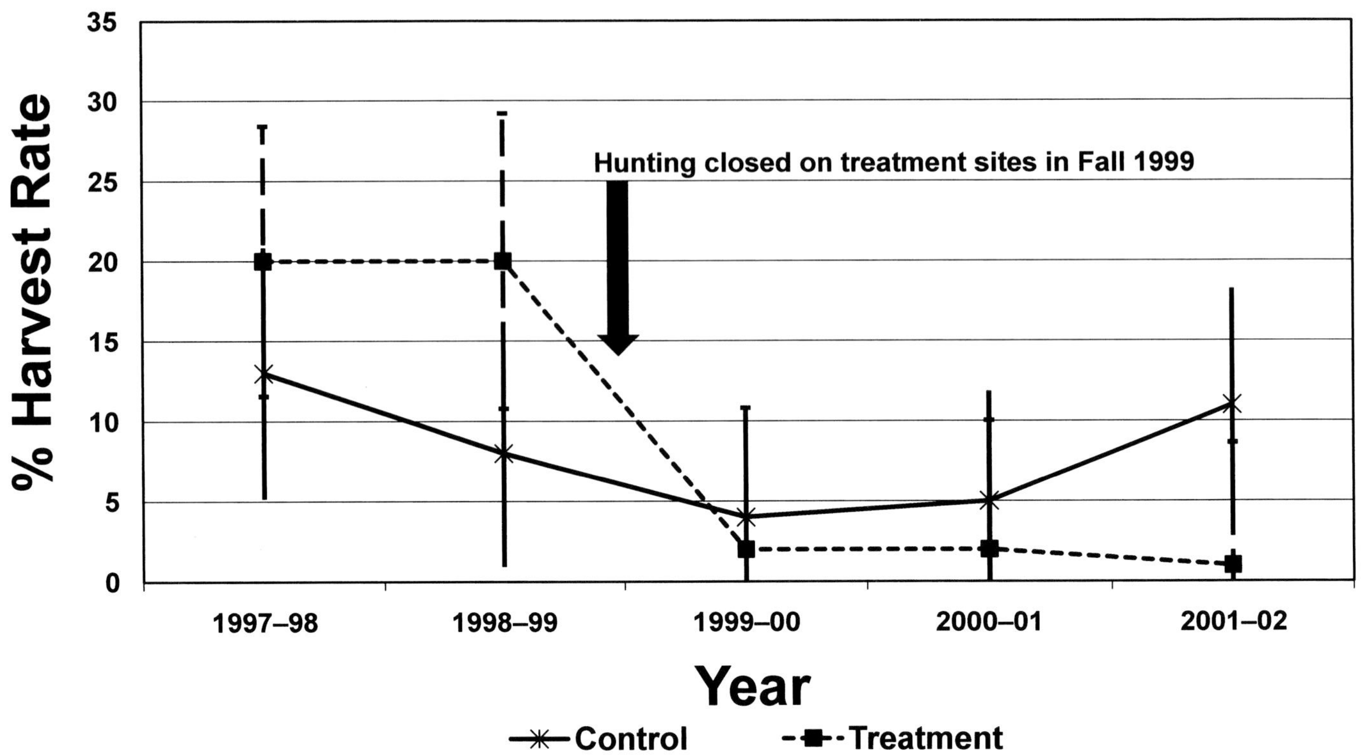

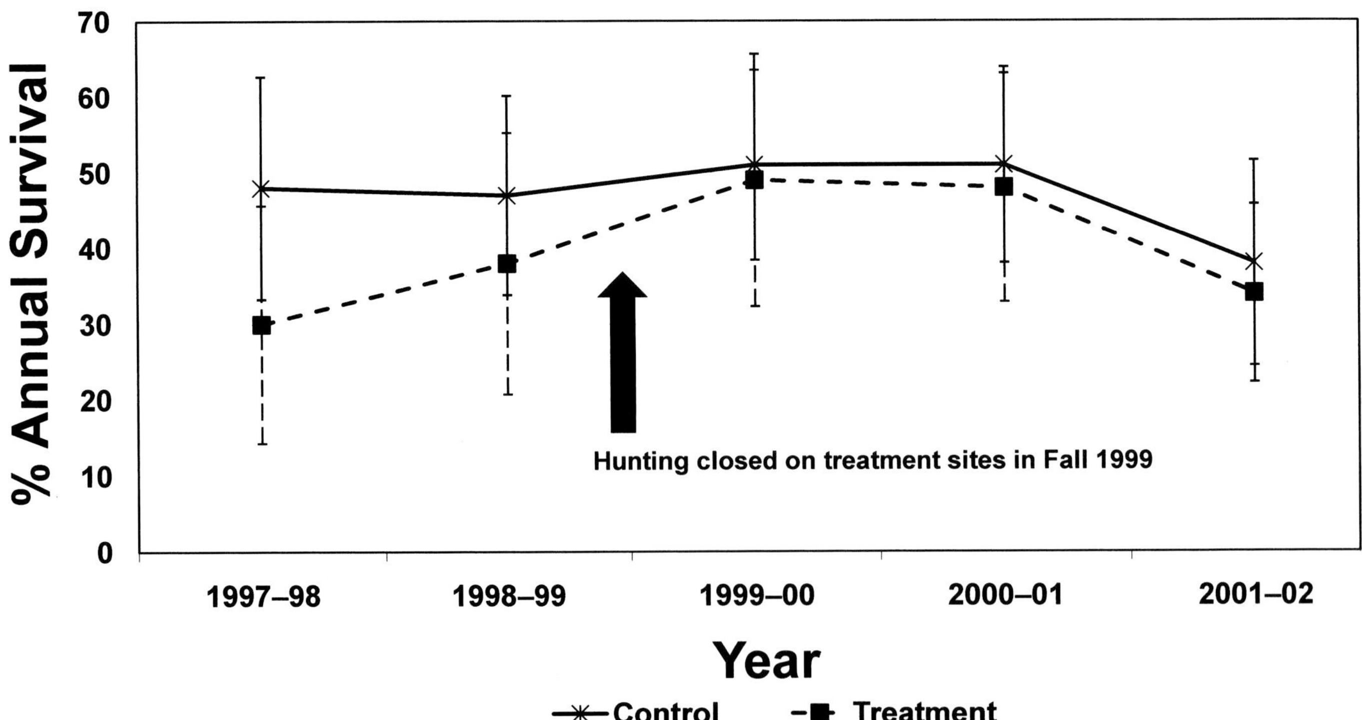

Figure 5.3

Top: Ruffed grouse harvest rates on control and treatment sites in the Appalachians, 1997–2002.

Bottom: Ruffed grouse annual survival in treatment and control sites in the southern and central Appalachian region, USA 1997–2002. Treatment sites were open to hunting from 1996–1998 (Phase I) and closed to hunting from 1999–2001 (Phase II). Control sites were open to hunting every year.

to detect a change in survival on the three treatment sites after closing hunting. Importantly, annual harvest rates declined on control areas from 1997 to 2000 and increased to the end of the study (Figure 5.3). The low harvest rates, particularly on control areas, experienced during this study reduced our ability to detect effects that might have occurred.

Although we believe regulated sport harvest did not have a direct impact on ruffed grouse survival, there is evidence that disturbance from hunting (and other activities) influenced habitat selection and home range size of ruffed grouse in the Appalachian region (Whitaker 2003, Chapter 10). Ruffed grouse (regardless of sex and age classes) made greater use of clearcuts and mesic bottomlands and had smaller home ranges in the absence of hunting. We believe this type of disturbance deserves consideration in the development of ruffed grouse hunting regulations and land management.

6

Predation

George B. Bumann, Patrick K. Devers,
Dean F. Stauffer, and Gary W. Norman

Hunters periodically cite increases in predator observations as being related to higher predation on game bird populations. Although anecdotal evidence may hint at such patterns for ruffed grouse, there is little information to evaluate the validity of such trends in the Appalachians. Concurrent estimates of species-specific predator abundance and predation rates on ruffed grouse are crucial to evaluating this hypothesis. This study was the first large-scale effort to quantify and relate predator abundance to predation on ruffed grouse in the Appalachians.

Periodic grouse population declines in the Great Lakes region and southern Canada have been associated with invasions by hawks and owls, particularly northern goshawks and great horned owls (Bent 1937, Lack 1954, Grange 1948). Low abundance of goshawks, the lack of predator invasions, and different climatic conditions likely account for different predation patterns in the Appalachians. We hypothesized that raptors (birds of prey), rather than mammals, would represent the leading cause of predator-related grouse deaths. (Figure 6.1 in color section) This was borne out by the information presented in Chapter 5—approximately 44% of all grouse deaths were attributed to raptor predation whereas about 26% was due to mammals; an additional 14% of the mortality was predator-related, but we were not able to determine the predator responsible. The Appalachian Mountain range is a prominent migratory route and summer breeding habitat for eastern raptors. Raptor nesting and migration may coincide with annual increases in avian predation on ruffed grouse. We also believed that grouse behavior associated with brood dispersal and nesting periods would result in higher predation mortality. We compared data collected at multiple sites throughout the region on a year, month, and site basis to elucidate these patterns of predation and predator abundance as part of the ACGRP.

Figure 6.1

Common predators of ruffed grouse in the Appalachian Mountains. Clockwise from top left: Cooper's hawk, great horned owl, bobcat, and broad-winged hawk.
Photos: istockphoto.com

Methods

The methods for grouse capture and monitoring have been described previously. In addition to the standard observational procedures used for grouse, we also monitored predator abundance continuously on all study sites. Avian and mammalian predator sightings were recorded at each study site by multiple observers during routine field activity. Each day they were in the field, observers noted the date, time-in-the-field, distance traveled, predator species seen, and the number of individual predators counted. We summarized this information for each site, month and year; rates of predator sightings were calculated on a per-hour basis. Species-level predator sightings, and individual counts, were recorded between February 21, 1997 and December 31, 2000.

We calculated the monthly predation rates for grouse as the number of deaths due to predation divided by the number of grouse alive at the start of each month for each study site, year, and month. Predation rates used in final analyses were categorized as avian, mammal, or unknown predator.

Results and Discussion

The information presented in this chapter comes from following 1,768 ruffed grouse captured on the 10 study sites between September 7, 1995 and December 4, 2000. Three hundred and eleven grouse died prior to the end of the seven-day conditioning period and were eliminated from this analysis. The remaining 1,331 ruffed grouse were used to estimate predation rates.

Between February 21, 1997 and December 31, 2000, predator observations were logged by more than 120 observers during 43,994 hours of field study and 345,553 km (about 215,000 miles) of travel. These data cumulatively accounted for 6,581 field days of observation and documented 4,281 separate predator sightings across the study areas. Avian predators tallied in this study included red-tailed hawk (985), unidentified hawks (968), broad-winged hawk (516), Cooper's hawk (335), red-shouldered hawk (315), barred owl (154), golden eagle (123), bald eagle (42), unidentified owls (33), great horned owl (21), and northern goshawk (13). Mammalian predators recorded were domestic dog (498), bobcat (90), raccoon (41), coyote (34), red fox (22), fisher (22), gray fox (21), house cat (17), weasel (15), mink (10), opossum (3), and striped skunk (3). Cause-specific predation rates varied across the study sites (Table 6.1). Avian predation was highest in the northern sites, and accounted for about 92% of all predation mortalities at the Rhode Island study site. Mammalian predation rates become higher as we moved south along the Appalachians, and were highest at the North Carolina study site at 66%.

Apparent abundance of avian predators varied with the year. The highest rate of avian

Table 6.1. Cause-specific (% of total mortalities) predation on ruffed grouse on the Appalachian Cooperative Grouse Research Project, summarized across region and site, 1996–2000.

Area	Avian % (*n*)	Mammal % (*n*)	Unknown % (*n*)	Total (*n*)
NORTHERN	59.5 (47)	22.8 (18)	17.7 (14)	79
ALLEGHANY PLATEAU	52.1 (101)	28.9 (56)	19.1 (37)	194
RIDGE AND VALLEY	47.2 (143)	41.9 (127)	10.9 (33)	303
KY1	59.6 (28)	19.1 (9)	21.3 (10)	47
MD1	44.8 (30)	38.8 (26)	16.4 (11)	67
NC1	29.5 (13)	65.9 (29)	4.5 (2)	44
PA1	53.7 (36)	25.4 (17)	20.9 (14)	67
RI1	91.7 (11)	8.3 (1)	0 (0)	12
VA1	74.3 (26)	20.0 (7)	5.7 (2)	35
VA2	50.0 (29)	41.4 (24)	8.6 (5)	58
VA3	35.7 (30)	40.5 (34)	23.8 (20)	84
WV1	53.8 (43)	26.3 (21)	20.0 (16)	80
WV2	57.9 (45)	40.2 (33)	4.9 (4)	82
Total	50.5 (291)	34.9 (201)	14.6 (84)	576

predator sightings/hour was in 2000, with about 0.10 sightings per hour. The lowest rate of raptor observation was in 1998 (0.06/hour); in 1997 and 1999, we detected 0.08 raptors each hour. Mammalian predators (excluding dogs) were observed much less frequently than raptors, with about one mammalian predator being observed for every 13 avian predators detected. This may result in part because mammalian predators tend to be nocturnal, and were less likely to be detected by field personnel during the day. Elsewhere, (Bumann and Stauffer 2002) we showed that much of the predation attributed to mammals might in fact be the result of mammals scavenging on birds initially killed by raptors. Based on their relatively low detectability, and the questionable role of mammals as predators of adult grouse (Bumann and Stauffer 2002), we chose to compare predation rates to avian predator indices only.

Overall frequencies of raptor observations peaked during the late summer and early autumn. July had the highest overall frequency of raptor observation (0.11 observations/hour). Red-tailed hawks were the most frequently identified raptor for all months (0.013 observations/hour, range = 0.002–0.042) except August, where they were slightly exceeded by broad-winged hawks (Figure 6.2B). Unidentified raptor sightings peaked during the months of July and August, and exceeded red-tails from May to October (Figure 6.2C). Because broad-winged hawks "can unquestionably claim the dubious distinction of being the most misidentified of our local birds" (Bent 1937:252), the latter category may contain a large number of broad-winged hawks. Broad-winged hawks were still the second most frequently identified raptor (0.001 observations/hour, range = 0.001–0.031; peaking during April and July) and exceeded red-tail hawk observations between July and September. Cooper's hawks (0.001 observations/hour, range = 0.001–0.021; Figure 6.2F) were the third most frequently observed hawk, and peaked in number from August through October, with a small increase in sightings during April. Red-shouldered hawks (0.002 observations/hour, range = 0.001–0.015; Figure 6.2D) were observed in roughly equal numbers between January and June and infrequently during July through December. Owls (Figure 6.2E), eagles and goshawks were observed in much lower numbers than the aforementioned species. With the exception of owls, these species were not included in the anal-

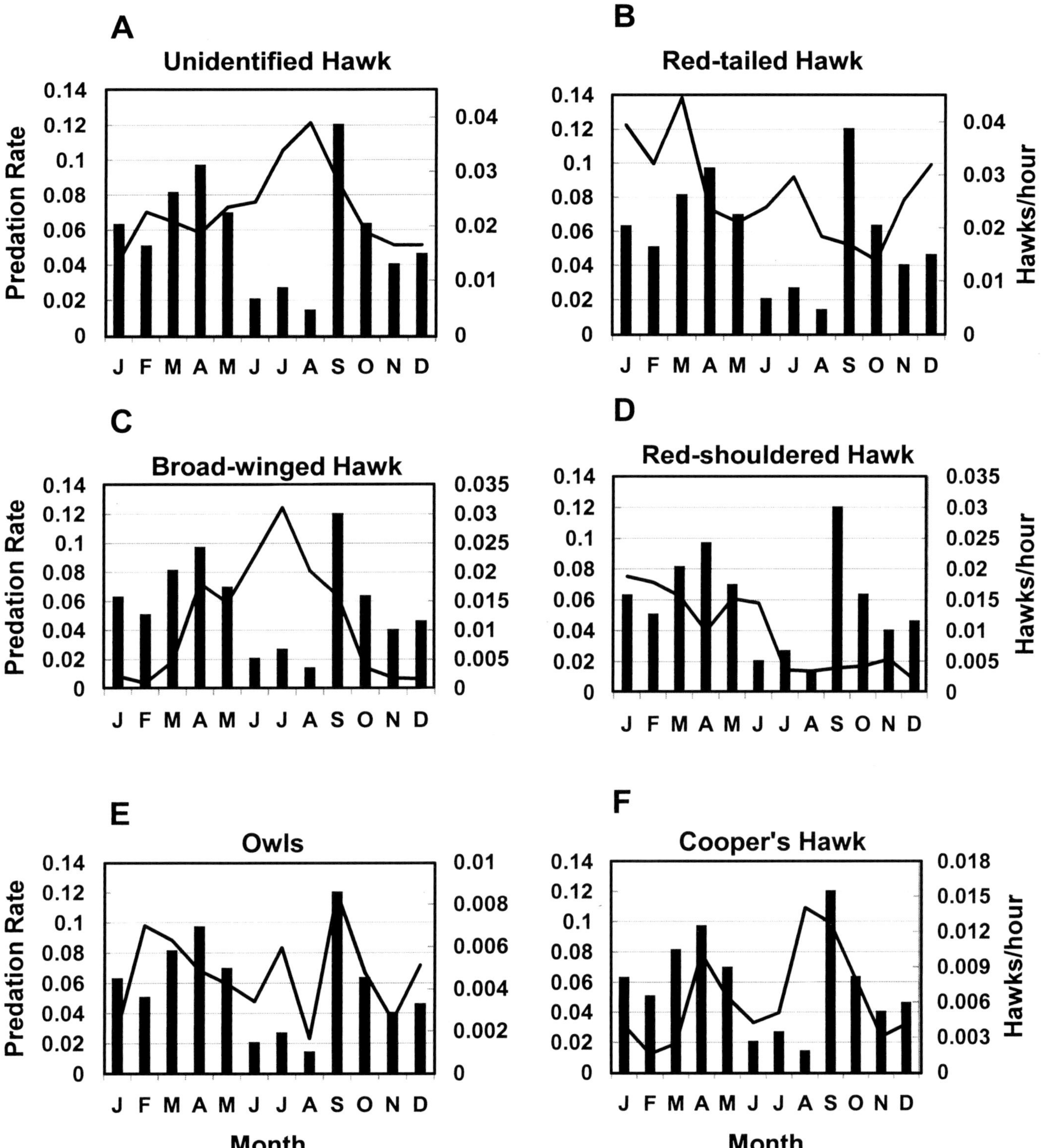

Figure 6.2

Monthly predation rate (proportion of deaths attributed to raptors) on ruffed grouse (bars) and frequency of raptors observed by month (line), pooled across sites and years in the Appalachians between 1997 and 2000.

ysis due to their questionable role as grouse predators (eagles) or small sample size (goshawks).

Five hundred seventy six grouse deaths (82.8% of mortality) were attributed to predation during the period of this analysis. Fifty percent (n = 291) of predation events were attributed to raptors, 35% to mammals (n = 201), and 15% (n = 84) to unidentified predators. The months of highest predation on grouse were during September (12% predation rate) and April (10% predation rate; Figure 6.2). Months of lowest depredation were August (1%), June (2%) and July (3%).

There was no strong relation between monthly predation rates and abundance of red-tailed, broad-winged and red-shouldered hawks, or with owls. However, there appeared to be a possible relation between the abundance of Cooper's hawks and grouse predation rate (Fig. 6.2). In all months except August, the predation rate closely tracked Cooper's hawk abundance; and, excluding August, this relationship was highly statistically significant. August was the only month when large numbers of uncollared, adult-sized grouse entered the population; trapping and collaring activities began in September each year. As a result, predation occurring in August likely was undetected or underestimated.

Patterns of cause-specific predation for ruffed grouse in this study were similar to those of other investigations (Bump et al. 1947, Gullion 1970, Clark 2000). Predator sightings did increase during the four years of the study, but were probably due to increased familiarity of the field staff with research sites and predator habits and identification rather than predator invasions. If invasions of predators do occur, and anecdotal evidence for this is lacking, they cycle at an interval of more than four years; patterns to support invasions were not captured by this study. Sighting frequencies for the raptor species most likely to prey on grouse appeared to coincide with seasonal migration and breeding activity of the raptors (Sauer et al. 2004). Summer counts of raptors were associated with fledging of nesting species and August migration of broad-winged hawks.

Each of the raptor species considered are known to prey on ruffed grouse (Bent 1937). Great horned owls and red-tailed hawks have impacted grouse numbers in northern populations (Bump et al. 1947, Rusch et al. 1972, Keith et al. 1977). Bump et al. (1947) also found a negative relation between the occurrence of ruffed grouse remains in owl pellets and grouse survival in New York. "Small" hawks such as broad-winged hawks are often ruled out as grouse predators, yet Rusch and Doerr (1972) found ruffed grouse to be the largest volumetric component of broad-winged hawk diets in Alberta, Canada.

Seasonal predation rates reflect findings from studies conducted elsewhere in grouse range (Marshal and Gullion 1965, Rusch and Keith 1971, Gormley 1996). Predation was reported to be most common during the spring and autumn in these studies. Spring predation may have been related to raptor migration and heightened activity of grouse during breeding and egg laying. Mueller and Berger (1992) demonstrated that migrating raptors in Wisconsin attacked lures at higher frequencies and of larger size in spring than in fall. At this time, male grouse are actively territorial, females are moving greater distances to visit males, and both sexes are engaging in courtship behavior (Gullion and Marshal 1968, Doerr et al. 1974, Boag 1976, Maxson 1977). Peak counts of ruffed grouse observed along roads and travel routes were coincident with periods of high predation on grouse in spring (ACGRP unpublished data). A peak in autumn predation probably reflected an abundance of young, inexperienced grouse in the population available to migrating raptors (Bergerud and Gratson 1988).

Correlations between predation rates on grouse and the abundance of owls and Cooper's hawks in the ACGRP suggest that these species may be important grouse predators in the Appalachian region, especially in fall when peaks in abundance coincided with predation rates. Monthly comparisons of predation rates and buteo sightings however, failed to show any relation in this study. Though it is probable that each predator species hunts grouse in the Appalachians, Cooper's hawks and owls may occupy the role of principle grouse predators in the Appalachians. Cooper's hawks are the most common, large Accipiter hawk in the Appalachian forests and may be the most prominent of all grouse predators in the Mid-Atlantic.

Although these results are based simply on correlations between grouse survival and raptor numbers, it does appear that we have some indication of predation patterns on ruffed grouse. The synchronicity in time and space for certain predators and grouse death is evident, but the true nature of this relation cannot be identified without a true experimental approach. These patterns, however, may provide evidence for previous hypotheses that remained unsupported by empirical evidence.

Large fluctuations in predation trends were not observed in the central Appalachians during the years of this study. Combined with anecdotal evidence, this would suggest that predation on ruffed grouse is relatively constant among years. Predation on grouse in the Appalachians does appear to be influenced by seasonal factors, with migration of raptors and breeding activity of grouse perhaps being foremost. The role of grouse courtship behavior, the influx of juvenile grouse into the population, and the hunting success of resident versus migrant predators are deserving of further research.

7 Food Habits and Nutrition

Robert Long, John W. Edwards, Roy L. Kirkpatrick, and Aaron Proctor

Food is essential for any animal; the amount and quality of food has the potential to affect many aspects of grouse ecology including their behavior, activity levels, body condition, survival, and reproduction. A growing body of research provides compelling evidence that nutrition may be an important factor influencing grouse populations in the Appalachians. In this chapter, we review the food habits and nutritional ecology of ruffed grouse and discuss the possible role that food may have in affecting productivity and population dynamics in the Appalachians.

Pre-breeding Food Habits and Condition

Throughout this chapter we will rely extensively on the research of the Appalachian Cooperative Grouse Research Project and particularly studies conducted by Long (2007) and Proctor (2010) that investigated pre-breeding food habits, nutrition, and body condition of ruffed grouse and potential effects on productivity in the Appalachians. Long (2007) collected 352 ruffed grouse from eight ACGRP study areas in the Appalachian region and 80 ruffed grouse from Michigan, Wisconsin, and Minnesota for comparison. Collections occurred in three consecutive years from 2000–2002 and were timed to occur nine weeks prior to peak hatching date for each site. Collection dates were approximately two to three weeks prior to the initiation of egg laying. Crop contents of collected birds were examined and separated into 11 forage categories based on structural and nutritional differences: (1) herbaceous leaves and flowers; (2) evergreen leaves; (3) deciduous leaves; (4) ferns; (5) buds and twigs; (6) oak and beech fruits; (7) other hard fruits; (8) soft fruits; (9) catkins; (10) aspen flower buds; and, (11) animal matter. An Importance Value was developed (IV = [% of aggregate crop contents/100 + occurrence (%) / 100] / 2) to reflect and assess the relative importance of forage classes and individual foods on a scale of 0 to 1. Percent carcass fat is generally considered the most accurate measure of body condition and was calculated using standard proximate analysis techniques. Reproductive data were gathered at ACGRP study sites as per protocols outlined in Chapter 3. Further detail regarding the methods and data analyses used in this research can be found in Long (2007).

Food Habits of Appalachian Ruffed Grouse

The diversity of the Appalachian grouse diet is well documented. Grouse are primarily vegetarians with a few exceptions, typically consuming a variety of leaves, buds, twigs, seeds, and fruits throughout the year. Food resources also vary substantially seasonally, annually, regionally, and locally. A thorough review of grouse food habits here would be unnecessarily lengthy and repetitive with Barber et al. (1989). Therefore, in this section we will focus on what we feel is the most nutritionally stressful and critical period for Appalachian grouse—winter and early spring. From spring green-up through autumn, food is generally abundant. Summer diets are composed of various herbaceous plants that grow in the forest understory, and chicks obtain much-needed protein from an insect-rich diet during their first few weeks. Ripe soft fruits such as blueberries and blackberries are also a major summer staple. As summer turns to fall, more plants begin to bear fruit and grouse will capitalize on the abundance of choices. They will readily consume acorns, beechnuts, greenbrier berries, grapes, sumac fruits, rose hips, hawthorn fruits, catkins, and a variety of herbaceous plants. Unfortunately, many of these foods disappear quickly. Most are also high on the preferred food lists of other wildlife, while others simply decompose or die back with the onset of cold weather and snow. The "time of plenty" soon becomes a "time of scarcity" for many Appalachian ruffed grouse.

Long (2007) found the food habits of Appalachian grouse in the winter and early spring to contrast sharply with those of northern populations (Figure 7.1). In the Lake States, Canada, and New England, high use of aspen flower buds in winter and early spring is well documented

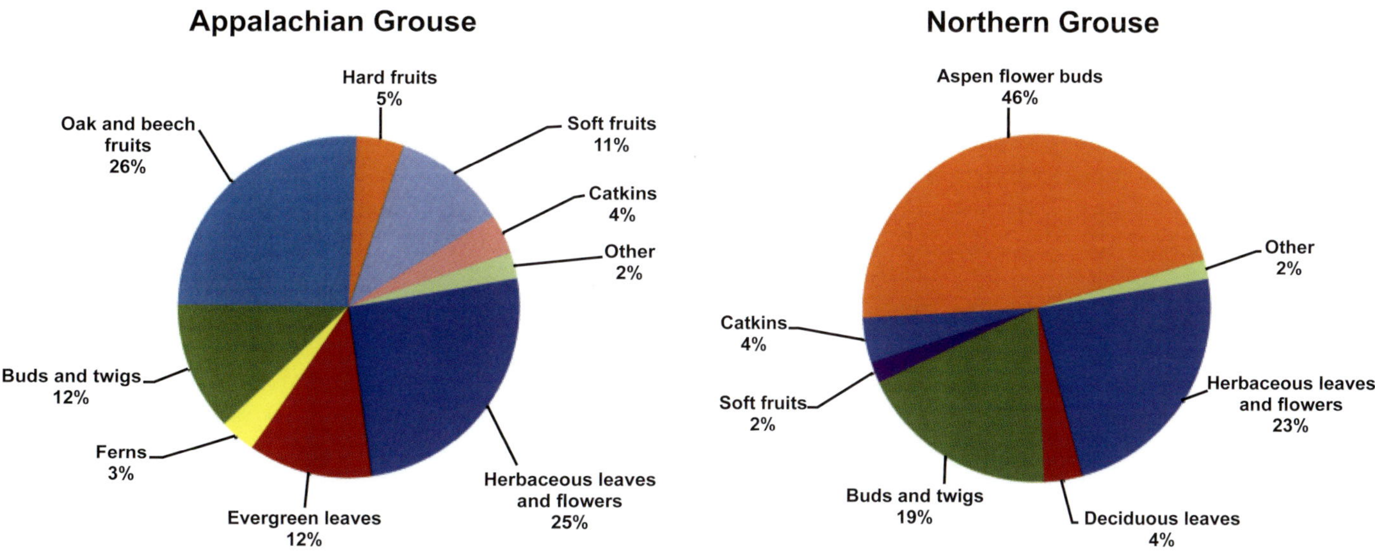

Figure 7.1

Percent of total mass of forage classes found in crop contents of northern grouse and Appalachian grouse collected in March to April 2000–2002.

(Darrow 1939, Bump et al. 1947, Stollberg et al. 1952, Svoboda and Gullion 1972, Doerr et al 1974, Woehr and Chambers 1975). Long (2007) also documented extensive use of aspen in Minnesota, Wisconsin, and Michigan in the pre-breeding period, with flower buds accounting for 46% of crop content mass (Figure 7.1). Dependence on aspen in the North should come as no surprise; throughout much of its range, aspen is a readily available and easily obtained food source of moderate nutritional value.

Research has shown that grouse can fill their crops rapidly when feeding on aspen, minimizing movements and risk of predation. Foraging sessions have been reported to average only 16–24 minutes (Svoboda and Gullion 1972, Doerr et al. 1974). But, unlike their northern counterparts that typically can make quick forays to collect aspen buds, grouse in the southern Appalachians need to be adaptable. Researchers at Virginia Tech estimated that ruffed grouse in the Appalachians needed to forage at least 100 minutes to meet energy demands, and they found radio-tagged grouse in Virginia to be active an average of five hours per day (Hewitt and Kirkpatrick 1996, Hewitt and Kirkpatrick 1997b). Grouse in the Appalachians feed on an amazing variety of foods, but it appears that resources can be widely scattered at times, and only the best habitats have abundant foods during the critical winter and early-spring period. Long et al. (2006) found 40 food items commonly occurring in crops of grouse collected in the Appalachians in March and early April (Table 7.1). However, most individual grouse had fed on only a few items prior to collections.

Grouse at ACGRP sites foraged more on herbaceous, green vegetation than was reported in previous studies (Long 2007). Although most previous food habits research analyzed crops from hunter-killed grouse, this study collected specimens outside the hunting season, at about the same time that some early-sprouting species were growing. Herbaceous leaves or flowers occurred in over 90% of Appalachian grouse crops, suggesting that recently emerged vegetation is an important nutritional resource that is highly sought during spring green-up. Flowers and leaves of coltsfoot and cinquefoil were among the most consumed herbaceous forages. The transition to more herbaceous plants might be expected if we consider the nutritional requirements for breeding grouse. Robbins (1981) estimated that protein requirements during egg laying are 175% greater than during winter in gallinaceous birds. Herbaceous leaves are the most protein-rich forage class available to laying grouse, containing 21–29% protein (Servello and Kirkpatrick 1987). However, annual variation in herbaceous food use was evident. For example, in Pennsylvania, herbaceous leaves made up 56% of crop contents in 2001, but they were relatively unimportant in 2000 when aspen flower buds, buds and twigs, and soft fruits were the most dominant forages. Although collections occurred during the same week each year, spring green-up was later in 2000, which may have limited the availability of "green" forages.

Although herbaceous plants were eaten regularly by grouse in this study, it is likely that few grouse had access to abundant herbaceous forage in January and February. Most forbs and flowers found in crops were among the earliest species to emerge in spring, such as cinquefoil and coltsfoot, and were in early stages of development. Previous studies have suggested that grouse in the Appalachians maintain themselves during winter on hard and soft mast, buds, twigs, or evergreen forages that they can find. Foods such as acorns, mountain laurel leaves, Christmas ferns, greenbrier berries and leaves, grapes, and birch and cherry buds are commonly reported winter foods (Stafford and Dimmick 1979, Seehorn et al. 1981, Servello and Kirkpatrick 1987). Servello and Kirkpatrick (1987) documented that herbaceous plants made up less than 20% of the diet during January and February in southeastern states with the majority of the diet consisting of leaves, buds, and twigs of woody species and fruits.

Evergreen leaves, primarily mountain laurel, greenbrier, dewberry, wintergreen, and trailing arbutus, are among the poorest quality forages but are consumed regularly in the Appalachians. Research suggests evergreen forage use peaks in February and can compose 30–45% of the diet in January and February (Smith 1977, Servello and Kirkpatrick 1987). On ACGRP sites in March and early April, evergreen leaves occurred in 36% of

Table 7.1. Mean Importance Values [IV= (aggregate % mass / 100 + % occurrence)/ 100) / 2] of forages from crops of ruffed grouse collected in March and April 2000–2002 at 8 study sites in the central and southern Appalachians. Only forages with IV > 0.05 in at least one year are presented. Abbreviations are l.= leaves, fl.= flowers, c.= catkins, bt.= buds and twigs, and fr.= fruit.

	Site							
Forage	**PA**	**MD**	**WV1**	**WV2**	**VA1**	**VA3**	**KY**	**NC**
Animal matter	0.06	0.05		0.04			0.02	0.02
Aspen fl.	0.10							
Avens l.	0.14	0.15			0.04		0.03	0.04
Azalea l.								0.02
Beech fr.	0.02		0.05				0.31	
Birdsfoot-trefoil l.		0.20			0.05		0.03	
Black birch bt.	0.07	0.11	0.05		0.02			0.06
Black birch c.	0.10	0.03	0.07		0.03			0.05
Blueberry / huckleberry bt.	0.22		0.04	0.19	0.09	0.11		0.07
Cherry bt.	0.04		0.02					
Cherry fr.		0.06						
Christmas fern l.		0.14	0.06	0.06	0.14	0.21	0.11	0.20
Cinquefoil l.)	0.15	0.21	0.13	0.32	0.18	0.24		0.19
Clover l.			0.15	0.07	0.05	0.13		0.18
Coltsfoot fl.	0.16	0.03	0.06	0.07	0.12	0.23		
Dewberry l.	0.04		0.10		0.02			
Grape fr.	0.02							
Greenbrier fr.	0.02		0.11	0.08	0.07	0.04		0.07
Greenbrier l.			0.06	0.05	0.12	0.04	0.08	0.08
Hawkweed l.	0.04		0.14	0.14	0.06			0.05
Hornbeam c.					0.03			
Maple fr.	0.05	0.06		0.16	0.02	0.04		
Mountain laurel bt.			0.10	0.09	0.04	0.04		0.02
Mountain laurel l.	0.05		0.23	0.23	0.16	0.14		0.23
Multiflora rose l.			0.06		0.03		0.10	0.03
Oak fr.	0.22	0.10	0.04	0.27	0.19	0.35		0.09
Partridgeberry l.	0.07			0.05	0.10		0.03	
Ragwort l.						0.04	0.03	0.04
Serviceberry bt.	0.08	0.03	0.03	0.02	0.09	0.07		0.06
Sorrel l.		0.05						0.05
Strawberry l.		0.07				0.02		0.09
Sumac fr.	0.12	0.02					0.03	
Trailing arbutus l.	0.04			0.04	0.06	0.03		0.02
Viburnum spp. fr.					0.04			
Wintergreen fr.	0.03		0.06	0.05	0.07			
Wintergreen l.	0.03							
Witchhazel bt.			0.02	0.02	0.04			
Witch-hazel fr.			0.02	0.16	0.03			
Wood fern l.	0.07	0.23	0.15		0.04	0.04	0.02	0.07
Yellow Birch c.			0.05					

crops and accounted for 12% of the crop contents (Figure 7.1). Based on Importance Values (calculated using the percent occurrence and percent of crop contents) mountain laurel leaves were the most important forage at the WV1 and NC study sites but were also very important at the WV2, VA1, and VA3 sites (Table 7.1). Mountain laurel consumption was inversely related to hard mast use on most ACGRP sites. For example, at WV2 in 2000, evergreen leaves composed only 4% of crop contents when acorns were abundant (65% of crop contents), but in 2002 acorns made up only 5% of contents and evergreen leaves were consumed by nearly every grouse and accounted for 34% of crop contents. It appears that local mast failures may force some Appalachian grouse to consume substantial quantities of evergreen forages.

Acorns and beechnuts are an important component of the pre-breeding diet of Appalachian grouse (Norman and Kirkpatrick 1984, Servello and Kirkpatrick 1987, Servello and Kirkpatrick 1988, Long 2007). The persistence of these forages through March and early April on some ACGRP sites is somewhat surprising, though. Acorns and beechnuts made up a large portion of the aggregate diet in the Appalachians while being found in a small percentage of crops, suggesting acorns and beechnuts are consumed in large quantities when found (Long 2007). Masting patterns of oaks are highly variable (Koenig and Knops 2002) and our study suggested that mast availability was annually absent or spotty on many sites. Acorns were not found from crop contents at PA and VA3 in 2000 and 2001, but made up 78% and 90% of crop contents in 2002, respectively. On the WV2 site, acorns composed 65% of contents in 2000, and then decreased to 12% in 2001, and 5% in 2002. Beechnuts, though only found to be important in Kentucky, showed similar patterns of variability, accounting for 97% of crop contents in one year while absent the other two years. Servello and Kirkpatrick (1988) noted a similar "boom or bust" pattern to acorn use and, presumably, availability during March and April. Acorns and beechnuts are among the most energy-rich forages available for grouse (Servello and Kirkpatrick 1987, 1989) and appear to be highly selected when available.

Soft mast is an important summer, fall, and early winter food source according to many accounts (Stafford and Dimmick 1979, Seehorn et al. 1981, Norman and Kirkpatrick 1984, Servello and Kirkpatrick 1987). Long (2007) found that soft mast was eaten less in early spring than at other times. Most soft fruits were probably eaten or decayed prior to the collection period, and fruits of greenbrier, grape, and sumac were the only fruits found in substantial quantities. Bud and twig use varied, but occurred in nearly 50% of all southern crops. Consistent with the previously cited studies, birch, cherry, serviceberry, blueberry, and huckleberry were among the most common species of buds eaten. Buds and twigs are a low-energy, high-fiber food source that is readily available when other, more nutritious, species are absent.

The Importance of Nutrition

Food habits studies are limited in that they only allow researchers to document what foods grouse consume. Although that information is undoubtedly interesting and important, it is equally valuable to understand how a grouse utilizes that food and what consequences it may face if it does not obtain adequate nutrition. Although direct mortality of ruffed grouse due to starvation appears to be exceptionally rare, a fair amount of evidence has surfaced that suggests substantial indirect impacts of food limitations in the Southeast. Norman and Kirkpatrick (1984) were the first researchers to suggest that diet and body condition varied seasonally in the Appalachians, and expressed concern over possible effects on overwinter survival and reproductive success the following spring. Servello and Kirkpatrick (1987) followed with a thorough food habits and nutrition study that identified substantial regional differences in the nutritional ecology between the core grouse range and the Appalachians. They hypothesized,

> The likelihood of an abundance of high quality food throughout the winter in the southeastern states is low because fruit and herbaceous forb production from the summer and fall is probably insufficient to support grouse populations in winter, requiring a dietary shift to low qual-

ity evergreen leaves. Low grouse population densities in the Southeast may be, in part, the result of inadequate winter food supplies.

Subsequent research by Hewitt and Kirkpatrick (1997a, 1997b) provided further evidence that grouse in the Southeast might be food-limited. Based on these important studies, the ACGRP was compelled to explore the relations between diet and grouse populations more rigorously. Working with the ACGRP, Long (2007) found additional evidence that the pre-breeding diet of female grouse greatly influences body condition in the Appalachians.

The Diet-Condition Link. Servello and Kirkpatrick (1987, 1988) estimated the nutritional quality of many commonly consumed grouse foods. They found that most fall and winter forages, with the exception of some evergreen leaves, probably had sufficient metabolizable energy content to sustain Appalachian grouse, but questioned whether enough of these foods were available during the late-winter period. Although a field study to determine how much food a wild grouse eats does not seem feasible, we can glean information about diet adequacy by looking at body condition. If a grouse encounters nutritional stress and an energy deficit (i.e., using more energy than it takes in), body fat reserves will be the first source of energy utilized. However, once fat reserves are depleted, muscle catabolism will occur, essentially "burning" protein for energy in order to survive. Therefore, examination of grouse fat levels allows us to determine the relative quality of their diet.

The physiological condition of ruffed grouse, as indexed by the percent fat, appears to differ substantially among regions, local areas, and even individuals within the same area. Thomas et al. (1975) found that grouse in Ontario did not maintain high levels of fat reserves (approximately 6–9% carcass fat) and thought that regular feedings and aspects of northern grouse behavior that restricted heat loss were important in that region. Long (2007) also found that northern grouse generally did not contain large fat deposits. Grouse collected in the Michigan, Wisconsin, and Minnesota had 44% lower fat levels than Appalachian grouse on average (6.0% in the Lake States compared to 10.8% in the Appalachians). This finding might lead one to question why there is so much discussion about food limitations in Southeast. If grouse in the Southeast have more fat reserves than their northern counterparts that are generally found at higher densities, then it would appear on the surface that food resources are adequate in the Appalachians. The difference between the two regions likely lies in the vastly differing nutritional ecology.

In the central and northern part of the grouse's range, large amounts of moderately nutritional browse are typically available. Subsequently, there is probably no need to "stockpile" energy. In contrast, the fall, winter, and early spring diet of Appalachian grouse is exceptionally diverse overall and individual diets vary substantially depending on the local conditions (snow cover, cover type, masting patterns, etc.). Food is not evenly distributed and plentiful in the Appalachians. There are "times of plenty" and "times of scarcity." This is probably the reason that grouse in the Southeast accumulate large fat reserves if they are able—up to 40% carcass fat in one instance (Long 2007). Foods are probably most scarce immediately before the laying period when female grouse direct large amounts of energy and protein into egg formation, then incubate for 3–4 weeks with only minimal foraging time. Fat reserves have been shown to decline 30% between April and May, suggesting grouse may rapidly deplete body fat during the egg-laying period (Thomas et al. 1975). Long et al. (2006) found mean carcass fat levels in southern grouse to be highly variable depending on year and location, ranging from 5.6–27.0% for females and 4.1–19.0% for males. Values in this range have been reported for grouse collected in Virginia in December and January (Norman and Kirkpatrick 1984) and March (Servello and Kirkpatrick 1988). Females accumulated greater fat reserves than males in this study, consistent with previous accounts of ruffed grouse condition in Virginia (Servello and Kirkpatrick 1988, Norman and Kirkpatrick 1984). Low spring fat levels in males may be due to an increase in time spent near drumming locations and defending territories that may limit time spent foraging (Servello and Kirkpatrick 1988).

Long (2007) identified several foods that affected body condition (percent carcass fat) on a coarse level (i.e., aggregate crop contents compared to average percent fat for site) and at a finer level (i.e., crop contents of individual grouse compared to percent fat of that grouse). Strikingly evident was the relation between acorn and beechnut use and high fat levels. Females collected after consuming oak or beech mast contained an average of 71% more carcass fat than those not consuming hard mast (Figure 7.2). The effect of hard mast on male condition was even greater. Other researchers have reported anecdotal evidence to suggest that high acorn intake may contribute to increased fat reserves (Servello and Kirkpatrick 1988). Acorns are a highly digestible source of energy (Servello and Kirkpatrick 1989), and when abundant likely satisfy the dietary needs of grouse with minimal foraging times, which may also decrease exposure to predators.

We also confirmed previous researchers' observations that buds, twigs, and evergreen leaves tended to decrease fat reserves. These low-energy foods are often consumed during winter and early spring. Evergreen forages, particularly mountain laurel, contain toxic secondary compounds called phenols and are among the poorest-quality foods available to grouse in the Appalachians (Servello and Kirkpatrick 1987, Hewitt and Kirkpatrick 1997*a*). Hewitt and Kirkpatrick (1997*a*) found that grouse could maintain body mass with diets containing less than 20% evergreen leaves, but grouse consuming over 40% evergreen matter were unable to maintain body mass. Although evergreen leaves did not compose more than 40% of the composite diet at any ACGRP sites, the diets of individual grouse are largely unknown because individual crop contents only represent what the grouse consumed on the day of collection. However, 30 of 326 Appalachian crops contained more than 40% evergreen leaves and 13 crops contained over 75% evergreen leaves. If these grouse did not have access to other higher quality foods during late winter, excess consumption of secondary toxic compounds present in evergreen leaves may be affecting nearly 10% of the Appalachian grouse. Additionally, forbs and flowers were undoubtedly much less available in the weeks preceding spring green-up, leading one to assume that evergreen leaf consumption may be substantially higher during that critical period.

Potential Effects on Reproduction. As discussed, it is clear that ruffed grouse in the Appalachians feed on a variety of foods ranging from very high to very low quality during the critical late-winter period. Furthermore, the quality of that food greatly influences the physiological condition they can achieve prior to breeding. So, why is this important and what effect does it have on grouse numbers in the Southeast? Long (2007) found substantial evidence that the condition of breeding hens affects reproductive output. Although the ability to detect such a relation in the field is made difficult because of a variety of factors, on ACGRP sites that had hens with high fat reserves, we generally observed increased clutch size, nesting success, female success, hatching success, and chick survival.

Chick survival appears to be most strongly influenced by the condition of the laying hen. Whether a site had low, medium, or high fat levels accounted for 43% of the variation in chick survival to three weeks post-hatch (Long 2007). Weather has long been hypothesized to be a primary factor affecting early chick survival in ruffed grouse (Bump et al. 1947, Dorney and Kabat 1960), but female condition appeared to have a relatively greater influence on early chick survival in the ACGRP study. Chick survival on ACGRP sites with low fat levels in grouse was 61% and 45% lower than sites with moderate and high mean fat levels, respectively (Figure 7.3). The most probable explanation for this relation would be increased egg size or yolk size. Yolk content has been shown to increase in response to increased dietary energy and protein levels in other gallinaceous birds and previous studies of precocial birds found chick survival was positively related to egg size (Begin and Insko 1972, Gardner and Young 1972, Menge et al. 1979, Williams 1994). Thus, hens with greater body fat reserves likely lay eggs with more yolk reserves, which will contribute to survival of chicks in the critical first few days after hatching.

The effects of poor pre-breeding female condition do not appear to be compensated for as the

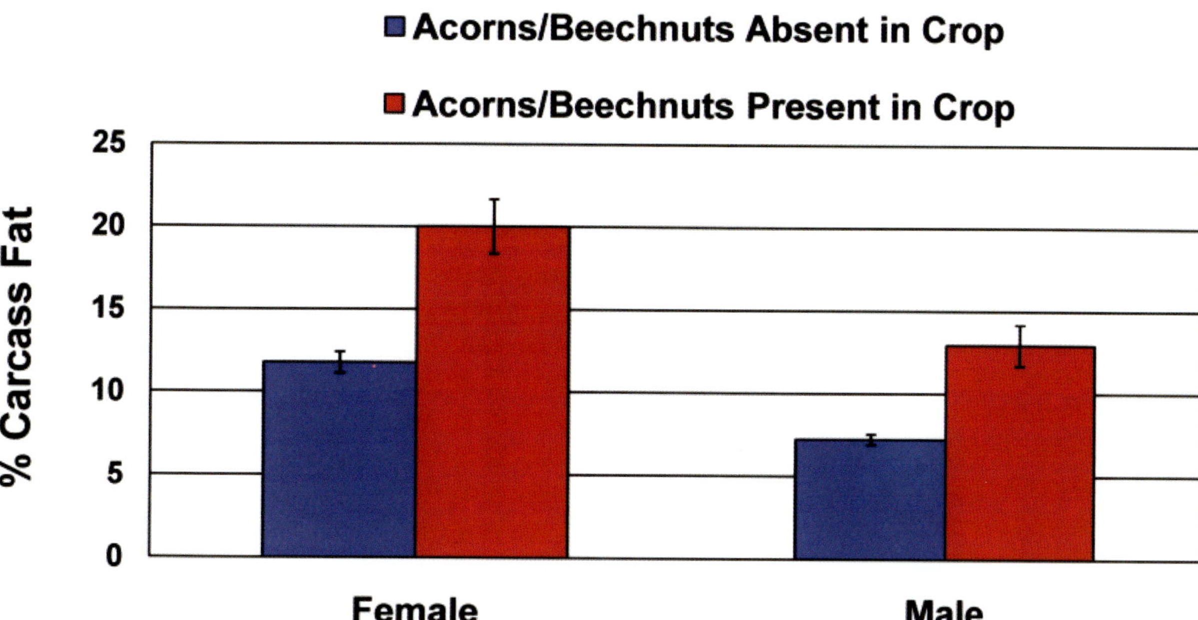

Figure 7.2

Percent carcass fat of Appalachian grouse collected with and without acorns or beechnuts in crop, 2000–2002. Vertical bars represent a 95% confidence interval.

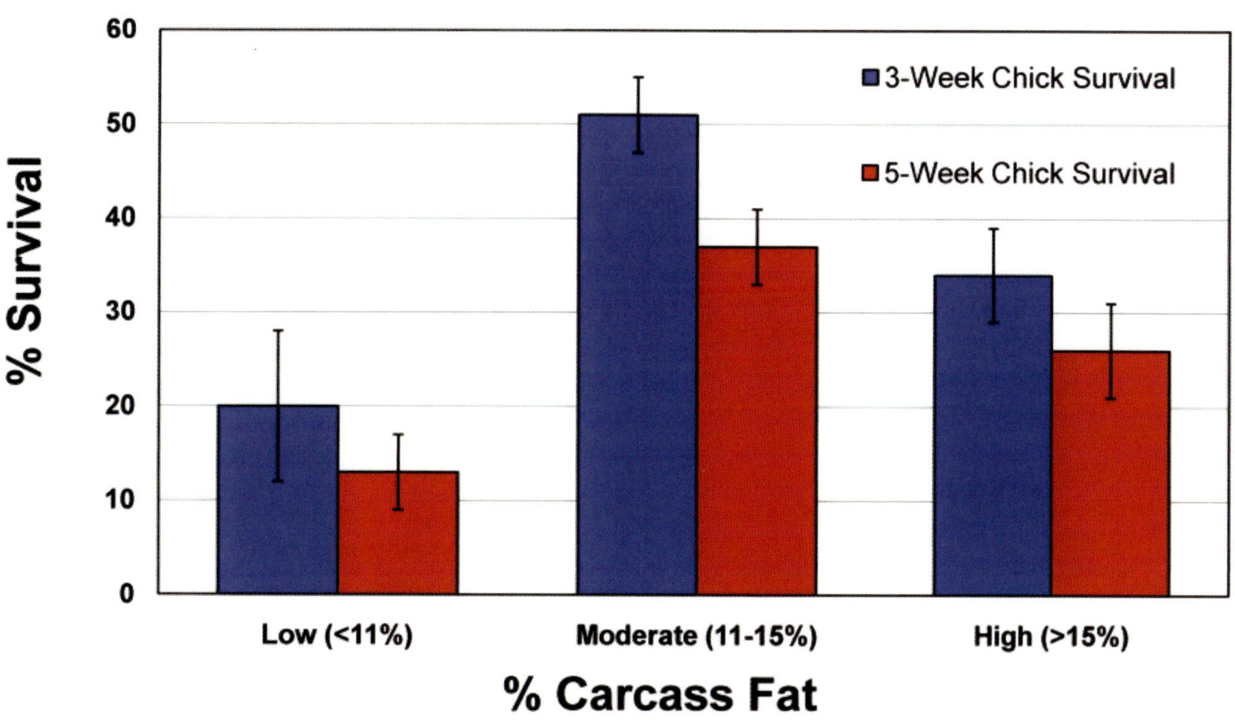

Figure 7.3

Chick survival of Appalachian Grouse on ACGRP study sites with low, moderate, and high levels of carcass fat, 2000–2002. Vertical bars represent 95% confidence intervals.

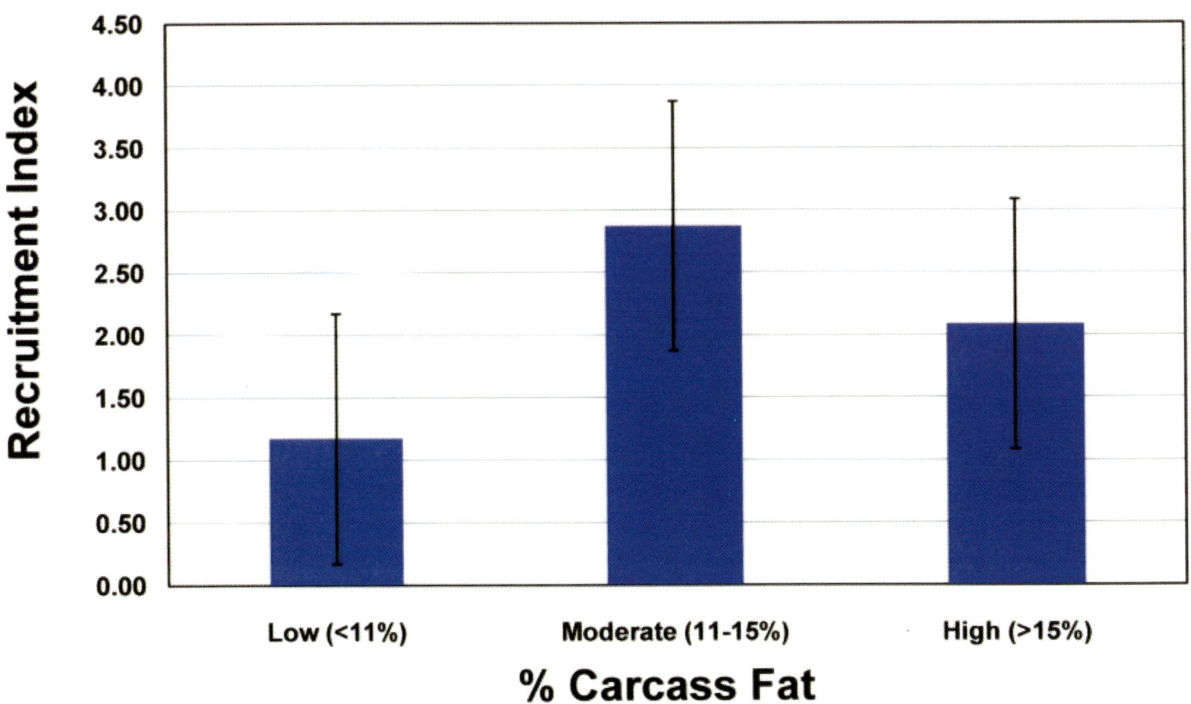

Figure 7.4

Figure 7.4. Recruitment Index (no. of chicks alive five weeks post-hatch per adult female) of Appalachian grouse on ACGRP sites with low, moderate, and high levels of carcass fat, 2000–2002. Vertical bars represent 95% confidence intervals.

brood matures, and fewer chicks were recruited at five weeks of age on sites with poor pre-breeding condition (Figure 7.4; Long 2007). The number of chicks alive for each hen at five weeks post-hatch on sites with grouse having low mean fat levels was 65% and 50% lower than sites with moderate and high mean fat levels, respectively. The ACGRP data suggest that survival and recruitment may be highest when grouse are in a moderate state of nutritional condition (11–15% carcass fat), and that productivity may actually slightly decline when grouse retain large amounts of body fat. This discrepancy may be explained in two ways. Grouse with abnormally large fat reserves may have been feeding exclusively on high-energy, low-protein food sources such as acorns. Large amounts of both energy *and* protein are needed for reproduction (Beckerton and Middleton 1982) and it is possible that protein deficiencies may account for lowered reproductive output. A more plausible explanation is that there is simply a threshold level of fat reserves that is needed for successful reproduction, and once this threshold is exceeded other factors become more influential than condition. It appears that when Appalachian grouse have less than 11% body fat, average reproductive output tends to decrease substantially, largely driven by poor chick survival. Conversely, grouse with average or above-average fat reserves likely reproduce adequately, on average, and chick survival is probably more dependent on weather, predation, and/or brood habitat constraints.

Further examination of the hen condition–chick survival hypothesis was conducted through experimental feeding trials of captive ruffed grouse (Proctor 2010). Female grouse were maintained on rations that differed in dietary energy and protein to directly quantify the influence of pre-breeding hen condition on reproduction. We examined 154 eggs from 103 clutches and found no relation between egg quality and hen condition. Moreover, egg composition remained consistent during the laying of the 10–12-egg clutch, regardless of hen condition. Although hen condition may possibly contribute to chick survival through increased yolk reserves, females ranging from 4–43% carcass fat in this study were able to produce clutches of similar quality and chick mass. An increase in egg size and chick mass associated with hen condition in other gallinaceous birds does not reflect structurally larger chicks, but ones having greater yolk reserves. An increase in yolk reserves should afford chicks an energetic advantage during the first few days post hatch—a period of high mortality in ruffed grouse. Our findings suggest that, in addition to hen condition, chick survival also may be influenced by post-hatch environmental factors (e.g., cold and inclement weather), habitat quality, and the hen's ability to avoid predation risks. It is possible that habitats providing high-quality foraging (i.e., producing hens with high carcass fat) may also independently possess excellent brood habitat, which, in part, could explain the differences in relations reported by the field study of Long (2007) and controlled lab study of Proctor (2010). The complex interplay among factors such as nutrition, environmental conditions, and predation makes it difficult to separate the influence of each in chick survival.

Numerous factors potentially affect local and regional grouse abundance and are addressed in other sections of this book. However, it appears that nutritional stress may influence grouse productivity in the Appalachians, and the degree of the stress is very dynamic, varying from year to year, location to location, and vegetation type to vegetation type. Consider an example from the Pennsylvania study site. Grouse were collected from two distinct portions of the study area that were approximately 16 km (10 miles) apart. The first—which we will refer to as Area A—had qualitatively excellent grouse habitat. The area was diverse, ranging from oak ridges with wild grapes and greenbrier thickets to young aspen stands interspersed with pockets of mature hardwoods that withstood a severe tornado in the mid-1980s. The second area—Area B—was a drier, more uniform habitat that was primarily dominated by mature chestnut oaks with a mountain laurel understory. Female grouse collected in Area A contained an average of 16% body fat with several females having at least 20% body fat. Males typically retain less fat, but still averaged 9% on Area A. Body condition was much poorer in the grouse collected in Area B; females had a scant 6% body fat and males were even lower at 5%. This example, although somewhat anecdotal, suggests that the available resources in a bird's home range can have dramatic influences on the nutrition and condition of breeding hens. The distribution and composition of vegetation types (and associated foods) is probably a key factor in determining the proportion of grouse that are in proper condition for breeding. Areas with a diversity of vegetation types that can supply high-quality foods even in the absence of unpredictable hard mast crops may hold the highest number of grouse, but in other areas where grouse are relying on spotty and unpredictable mast crops, lower densities of grouse may be more common. These higher quality areas might produce "surplus" grouse that can disperse to areas of lower quality where reproduction and survival are lower.

It also should be noted that food limitations would likely not only impact reproduction, but also have significant effects on other aspects of grouse ecology. Whitaker (2003) found home range sizes to be influenced by mast crops in oak-hickory forests. Increased movements due to scattered foods would logically put those grouse at a greater risk of predation as well. However, Devers (2005) did not find a relation between mast crops and survival on ACGRP sites. Unfortunately, many of the potential ramifications of food limitation may be difficult to decipher given the dynamic nature of field studies.

Our knowledge of the nutritional ecology of Appalachian ruffed grouse has grown substantially over the past three decades. As a result of multiple food habits, nutrition, and condition studies, several key points have emerged: (1) the diet of grouse inhabiting Appalachian oak forests differs substantially from grouse in the core (central and northern) part of their range; (2) grouse in the core portion of their range have significantly less fat reserves than Appalachian grouse, suggesting large differences in the nutritional ecology between regions; (3) hard mast, particularly acorns and beechnuts, appear to play a key role in increasing fat reserves in Appalachian grouse; (4) when hard mast is unavailable, grouse are forced to consume poorer quality forages, which often include evergreen leaves such as mountain laurel, depleting fat levels; and, (5) poor nutritional condition may influence reproductive success via lowered clutch size, nesting success, hatching success, and particularly chick survival.

Exactly why grouse populations in the central and southern Appalachians occur at lower densities will likely never be fully understood. However, this and previous studies provide compelling evidence that food limitation is an important piece of the puzzle. Nutritionally, the two regions are very different. In the north, grouse feed on a relatively constant supply of forages throughout the fall, winter, and early spring. Foods such as buds and catkins of aspen are of moderate nutritional value, but are abundant and widely distributed throughout the core of the grouse's range. Although grouse in the northern region may not accumulate large fat reserves, there is probably no need to; at any time they can move a short distance to meet their dietary and reproductive nutritional needs. However, Appalachian grouse do not have that luxury. The foods they consume are more variable in quality and availability. Hard mast, which appears to be the most selected forage of southern grouse, exhibits substantial annual and local variation. Grouse may have to traverse long distances to find these sometimes widely scattered pockets of mast-producing trees and forage in habitats where they are presumably more exposed to predators. Without mast, grouse typically eat a mixture of low-energy foods including evergreen leaves, ferns, and buds and twigs.

8

Roost Site Selection

John M. Tirpak, Darroch M. Whitaker, William M. Giuliano, and Dean F. Stauffer

Although the occasional male may drum on a moonlit night, ruffed grouse are typically active only between dawn and dusk. Therefore, it's not surprising that most studies of grouse usually have been restricted to daylight hours when the birds are most active and easily seen. However, many predators hunt by the cover of darkness, and temperatures are often coldest just before the dawn. Therefore, selection of nocturnal roost sites that provide both security and warmth may be an important consideration for grouse, and one with potentially grave consequences.

Because of the critical role roost site selection may have in survival, the ACGRP conducted two studies to identify the habitat features associated with roost sites in the Appalachians. The first investigated winter roost sites of individual birds at three sites in western Virginia (Whitaker and Stauffer 2003). The second examined the roosts used by broods during the summer in Pennsylvania (Tirpak et al. 2005). Together, they provide the first comprehensive view of the nocturnal habitat needs of grouse over the course of a year.

Methods

Studying ruffed grouse at night was not easy. While walking through a grouse covert by day has its own difficulties, wading through brambles and brush piles at night provides a whole new set of challenges. To locate roosting birds during the winter, researchers entered the woods just before dawn and used portable handheld receivers to home in on a bird's radio signal. As the researchers moved closer to the bird, the signal from the transmitter became louder. Ultimately, the researchers were close enough to flush the bird (typically less than five meters). Researchers then used piles of droppings to pinpoint exact roost locations (Figure 8.1).

A similar approach was used during the summer. However, because of concerns about disturb-

Figure 8.1

Piles of grouse droppings could be used to pinpoint roost locations. *Photo: Darroch M. Whitaker.*

ing chicks at night, every effort was made to not flush broods from their roosts. Instead, researchers used 200,000 and 1,000,000 candlepower portable spotlights to visually confirm brood locations. Nevertheless, finding a bird at night that is excellently camouflaged by day is not easy. Therefore, when researchers were unable to see the brood (but knew they were within a few meters of it), they marked the area rather than flushing the bird. They could then return to the site later and identify the exact roost location by the presence of fresh droppings.

After birds had left their roosts, researchers in both studies categorized roost position (i.e., on the ground or in a shrub or tree) and height (where possible), as well as the species, height, and diameter at breast height (dbh) of the nearest shrub or tree. Habitat structure was measured in 0.04-hectare (0.1-acre) circular plots around both roost sites and random plots located 50–100 meters away. This structure was then compared statistically between roost sites and random sites to assess habitat selection.

Winter Roost Sites

Winter is a hard time for ruffed grouse. Most deciduous trees have dropped their leaves and provide little concealing cover from predators. Additionally, temperatures in the winter commonly fall below the lower critical limit for grouse to maintain a positive heat balance. When the temperature is above 1.5°C (35°F), birds can maintain their body temperature through normal metabolic activity (Thompson and Fritzell 1988*a*). However, when the temperature drops below this threshold, birds are forced to increase their metabolic rate to avoid hypothermia. This requires energy, which necessitates longer foraging times and increases exposure to predators (Hewitt and Kirkpatrick 1997b). If birds fail to meet their increased energetic needs through daily food intake, they are forced to burn stored energy. For females, this stored energy is often fat and protein reserves that would have otherwise been invested in eggs. Therefore, when birds are forced to use this stored energy for heat, they enter the breeding season in poorer condition and have less energy to expend on reproduction (Servello and Kirkpatrick 1987). This lost energy may ultimately result in lower reproductive output the following spring (Beckerton and Middleton 1982, Chapter 7).

In northern regions, grouse gain an advantage from the deep snows that typically occur and are present throughout the long winter months. By roosting within the snow, birds are able to avoid the cold air and strong winds that would rob them of their precious heat and energy reserves (Gullion 1970). There, grouse are also relatively safe from predators, which cannot effectively detect birds hidden beneath the surface of the snow. Unfortunately for grouse in the Appalachians, conditions are rarely suitable for snow roosting. Snows are typically light and accumulations melt relatively quickly. Birds in the Appalachians are instead exposed to their predators and the elements for the majority of the winter.

Wherever snow is limited, grouse typically roost in dense evergreen vegetation (Whitaker and Stauffer 2003). These sites provide less protection than snow burrows, but more than leafless deciduous trees or barren open areas (Thompson and Fritzell 1988*b*). Birds in the Appalachians are often found roosting in evergreen habitats. However, unlike other regions with limited snowfall, coniferous stands dominated by pine, hemlock, or eastern red cedar trees are relatively uncommon. Instead, birds in the Appalachians rely on the dense evergreen shrub thickets (typically rhododendron or mountain laurel) common to this region. The greater availability of these thickets translates into increased use by birds. In Virginia, most nocturnal roosts occurred in evergreen vegetation, with more than half of these associated with evergreen shrubs (Whitaker and Stauffer 2003, Table 8.1). Although deciduous habitats were also used as roost sites, deciduous cover was more abundant than evergreen cover on these sites and only the latter received disproportionately higher use.

Table 8.1. Winter roost microsites used by ruffed grouse at three study sites in western Virginia (n = 90; 1998–2002).

Roost type	Count	% of Sites
Evergreen Shrub (ground)	13	14
Evergreen Shrub (above ground)	18	20
Conifer Tree (ground)	5	6
Conifer Tree (above ground)	13	14
Evergreen Total	*49*	*54*
Deciduous Shrub (ground)	5	6
Deciduous Shrub (above ground)	0	0
Deciduous Tree (ground)	19	21
Deciduous Tree (above ground)	10	11
Deciduous Total	*34*	*38*
Snow Burrow*	0	0
Open ground	7	8

*Grouse were regularly observed ground roosting in snow, but depth was never sufficient to allow birds to completely cover themselves.

Besides selecting the vegetation in which to roost, birds must also choose whether to roost on or above the ground. Wind speeds increase with height above ground, so roosting on the ground reduces wind-chill effects. However, a bird roosting on the forest floor is more susceptible to terrestrial-based predators like red foxes and weasels. The decision of where to roost therefore represents a tradeoff between predation risk and heat loss.

With any snow present (even not enough to be effectively covered), grouse will almost invariably roost on the ground, taking advantage of the lower wind speeds and insulating properties of snow. Of the 25 roosts Whitaker and Stauffer (2003) located when snow was present, 20 were on the ground. In contrast, when there was no snow, birds were fairly evenly split between roosting on the ground and in shrubs or trees above ground level (29 and 30 roosts, respectively). Roosting slightly above ground in an evergreen shrub may offer a good compromise between these two competing pressures and was the most commonly used evergreen roost site in Virginia.

This decision of roost selection was also influenced by the presence of rain during the night. On rainy evenings, birds preferred to be off the wet ground (six of nine roosts), which possibly creates an increased chilling effect. Also, rain may influence the quality of litter that birds may use for roosting and therefore affect roost position. Dropped oak leaves are much more resistant to decay and saturation than those of aspen or birch. Therefore, they remain relatively dry and lofted throughout the year and may develop litter layers up to 30 centimeters (12 in) deep. Ground roosts located in litter of these depths were characterized by bowl-like depressions where the grouse likely burrowed into the litter for thermal benefit and predator avoidance. On rainy nights, the litter layer becomes matted, which prevents birds from being able to burrow into it.

Where on a mountain a bird spends the night may be just as important as the type of vegetation in which it roosts. Because of the steep topography of the Appalachians, thermal inversions are common during the winter throughout this region. These thermal inversions occur on clear, still nights when cold air along the ridges slips down the side of the mountain and fills the valley with cold air. The warmer air in the valley is displaced by this colder air and rises, resulting in a pocket of cold air below the warm air above—the inverse of what is typically seen during the daytime (i.e., colder air at higher elevation). Birds seem to be aware of this meteorological phenomenon and

Table 8.2. Slope position of daytime and nighttime grouse locations at three study sites in western Virginia, 1998–2002.

Site		Number of Locations		
		Toe/ bottom	Midslope	Ridge
Virginia 1	Night	4	18	24
	Day	63	63	24
Virginia 2	Night	0	19	7
	Day	33	69	48
Virginia 3	Night	4	6	7
	Day	43	51	56
Total	Night	8	43	38
	Day	139	183	128

move from the warm valleys they occupy by day to the ridge tops at night to escape these plummeting temperatures (Table 8.2).

Habitat structure around winter roost sites tended to have greater coniferous cover and more coniferous trees than random sites (Table 8.3). In addition, stem density was greater at roost sites than random sites. In the absence of deciduous leaves and vegetation, dense stems may be an important component of both the vertical and horizontal cover that hides grouse from predators. Grouse are associated with high stem densities throughout their range during all seasons, and an abundance of small stems is likely one of the few common denominators between diurnal and nocturnal grouse habitat needs. Nevertheless, most grouse roosted in stands dominated by either relatively large pole-sized trees (12.5–28 cm [5–11 in] dbh; 29 roosts), or by sawtimber-sized trees (over 28 cm [11 in] dbh; 7 roosts) rather than saplings (less than 12.5 cm [5 in] dbh; 8 roosts). Within these mature stands, though, grouse selected areas with locally high stem densities. This may represent a compromise between the increased wind deflection of larger-diameter trees and the greater concealment of microsites in areas with higher stem densities.

Table 8.3. Habitat measurements from paired 400-m^2 (0.1 acre) sampling plots centered on 44 roost and control sites located 60 m away in a random direction at three study sites in western Virginia, 1998-2002. Values presented are the average for the variable.

Feature	Roost	Random
Deciduous canopy cover (%)	85	85
Coniferous canopy cover (%)	23	17
# 8–20 cm dbh conifers/ha	85	53
Conifer density (trees/ha)	115	78
Basal area (m^2/ha)	25	28
Stem density (stems/ha)	7453	5436
# stems within 3 m	52	34

Summer Brood Roosts

For most grouse, roost site selection in summer is not as critical as it is in winter. Nighttime temperatures are relatively warm, and dense vegetation provides concealment from both avian and mammalian predators. However, for young broods, selection of high quality nocturnal roost sites may be as important as selection of diurnal habitats. Young grouse are precocial and leave the nest soon after hatching. However, these chicks are equipped with only rudimentary feathers that are of limited use for either flight or thermoregulation (Johnsgard et al. 1989). Therefore, young are unable to fly to evade predators, and cool spring nights are often cold enough to thermally stress chicks. Not surprisingly, brood survival is lowest in the first few weeks after hatching (Larson et al. 2001, Smith et al. 2004, Chapter 4), and this high mortality is likely at least partly related to these physiological and developmental constraints. Selection of appropriate roost sites, however, may mediate some of these shortcomings.

Although roost site selection by broods in the summer and in the winter was likely driven by similar factors (i.e., predator avoidance and energy conservation), the relative position of roost sites differed between seasons. In contrast to winter, the decision to roost on or above the ground was not a function of precipitation, but rather reflected the developmental stage of the brood. The inability to fly during the first few weeks post-hatch prevents young birds from accessing roost sites off the ground. Nevertheless, even after young grouse are capable of flight at about 14 days, arboreal roosting is not common. Only five of 91 roosts located prior to the fifth week after hatching were off the ground (Tirpak et al. 2005). The poor ability of chicks to maintain their body temperature during these first five weeks prevents the young birds from roosting independently throughout the night. During this time, a hen will brood her young at night by covering them with her body and wings while on the ground. Around five weeks post-hatch, the young grouse presumably develop the ability to not only fly, but also effectively control their body temperature. From that point forth, grouse broods roosted exclusively in trees and shrubs off the ground. This pattern is probably a response to predation alone, as wind speeds and air temperatures during the summer are rarely severe enough to thermally stress birds once they are capable of thermoregulation. By roosting arboreally, grouse may be able to avoid ground-based predators, which are often more active during the night.

The specific habitat attributes at roost sites also differed between seasons, most likely as a result of changes in vegetation composition. As in winter, grouse selected roost sites in the summer characterized by dense vegetation. Both deciduous canopy cover and understory concealing cover were greater at brood roost sites than random sites (Table 8.4). However, the importance of evergreen vegetation for roost sites was much lower in summer than winter, and coniferous canopy cover did not differ between roost and random sites. This pattern is not surprising, as deciduous vegetation is fully developed by late spring and grouse are no longer limited to roosting in evergreen shrubs and trees—the only available cover in winter.

Grouse in both winter and summer roosted in areas with high stem densities. Although stem density did not differ significantly between brood roost and random sites, sapling stands were more abundant on the Pennsylvania site than on the Virginia sites used in the study of winter roost preferences. This increased abundance of sapling stands translated into increased use by broods and higher average stem densities at summer roost sites compared to winter locations. So, although stem density was not significantly greater at roost than random locations, it was greater at summer than winter roost sites. However, greater stem density at roost locations was also associated with lower apparent brood survival on the Pennsylvania site (Figure 8.2). Because stem density likely enhances survival by reducing detection of broods by predators, the association of greater stem density with lower brood survival observed by Tirpak et al. (2005) is surprising and probably not directly linked. Broods rely on arthropods, which they glean from understory vegetation, for food. Lower availability of arthropods in regenerating sapling habitats relative to mature forests may negatively affect survival of broods in these

Table 8.4. Habitat measurements at roost and random sites of ruffed grouse broods (n = 24) at the Pennsylvania study site. Values presented are the average for the variable.

	Roost	Random
Deciduous canopy cover (%)	81	69
Coniferous canopy cover (%)	4	2
Ground cover (%)	69	70
Coarse woody debris (%)	14	12
Basal area (m^2/ha)	14	15
Stem density (stems/ha)	18,671	16,678
Concealment cover – 5 m (%)	52	44
Concealment cover – 10 m (%)	71	64

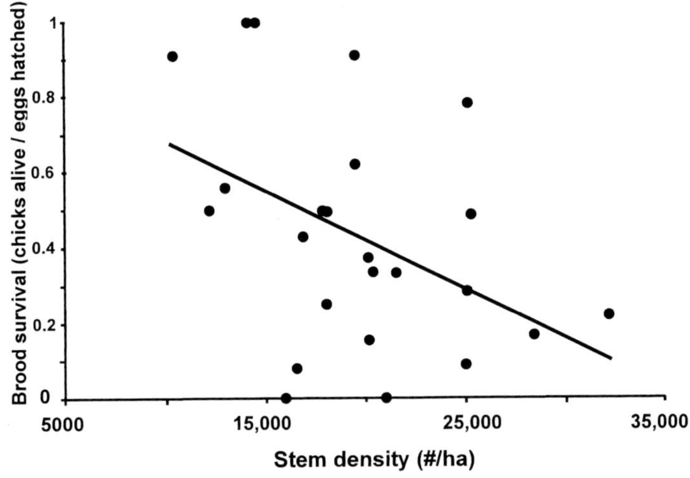

Figure 8.2

Relationship of chick survival within broods and stem density on PA1.

stands. Haulton et al. (2003) observed broods preferentially using mature forests with rich herbaceous understories and abundant arthropods over sapling stands with greater stem densities but with fewer insects. Sapling stands used by broods on the Pennsylvania site often lacked the abundant ground vegetation that has been associated with greater numbers of arthropods on other sites (Kimmel and Samuel 1978, Hollifield and Dimmick 1995). Therefore, broods may have been more secure from predators in sapling stands but may have experienced lower food availability.

Regardless of season, grouse roost in dense vegetation to reduce the effect of predators and inclement weather. In winter, this translates into increased use of evergreen habitats—mainly shrub thickets and young conifer stands. Although many managers specifically plant white pine for its perceived value as roosting cover, the relatively open growth form of this species reduces its value for this purpose, particularly when mature. Where winter roost habitat is limiting, conifers with denser foliage (e.g., eastern red cedar) could be planted as they provide a greater benefit. Alternatively, stands with high stem densities may improve escape and roosting cover for grouse in winter. Orienting clearcuts or heavy selective cuts so they straddle valleys and ridges would likely increase their use by grouse by providing secure travel corridors that connect foraging and roosting habitats.

Brood habitat management rarely considers the nocturnal habitat requirements of ruffed grouse broods during the early spring and summer. Traditional management practices to benefit broods (e.g., seeding logging roads, thinning, or burning forest stands) have focused on increasing herbaceous understory density and the abundance of arthropods that are the staple of the young chick's diet. By promoting understory vegetation, managers may increase concealing cover of the brood from ground-based predators but reduce overstory cover, which conceals birds from avian predators. Additionally, logging roads may serve as travel corridors for predators. Management prescriptions should recognize the importance of both nocturnal security and diurnal food availability. Incorporating seeded logging roads in early successional stands, or conducting heavy selective cuts or prescribed burning in mature forest stands may provide the necessary interspersion of these competing habitat-needs to promote brood survival.

9

Habitat Requirements

David A. Buehler, Craig A. Harper,
Darroch M. Whitaker, Benjamin A. Jones,
Carrie S. Dobey, Jennifer F. Kleitch,
Joy O'Keefe, Todd R. Fearer,
Eric Endrulat, John M. Tirpak,
and Dean F. Stauffer

Habitat Defined

Habitat for a particular species represents the physical and biotic resources (food, cover, water, and space) needed for populations of that species to perpetuate. Included in this definition is a consideration of the spatial configuration of those resources and the availability and use of those resources over time. Key life history requirements for ruffed grouse must be linked in space and time in such a way that an individual grouse can successfully move throughout the diurnal cycle, for example, from a nighttime roost site, to a foraging area or a loafing area, and back to a roost site, without undue risk of predation. Similarly, grouse must be able to link seasonal habitat requirements successfully. For example, nesting sites must be linked to brooding habitat so the successful hen can move her brood from the nest to appropriate brood cover without undue risk of predation. In general, the suitability of grouse habitat must be assessed at multiple spatial and temporal scales. In many cases, characteristics of individual habitat patches are not the only determinants of grouse habitat suitability. The interspersion and juxtaposition of these habitat patches are of at least equal importance in determining overall habitat quality.

Various, sometimes confusing, terms have been used to describe the habitat of a particular species. We use the term *habitat use* to indicate a descriptive account of the physical and biotic conditions present where a grouse is located. This can be characterized at different scales. The landscape scale includes the cover types and their spatial arrangement found where grouse are located. At the same time, habitat use can be described at a microhabitat scale, such as the number of saplings per acre found where a nest is located. *Habitat selection* refers to the process whereby an individual chooses the conditions that it occurs in. Habitat selection may occur across scales in a hierarchical fashion, including the geographic range (first order, as described by Johnson 1980), landscape or home range (second order), components within the home range (third order), and micro-site (fourth order). Microhabitat (vegetation characteristics such as stem density, etc.) would be considered part of third order selection. *Habitat availability* refers to the accessibility and procurability of various habitat components at any given temporal or spatial scale (Hall et al. 1997). *Habitat preference* refers to the end result of the habitat selection process, leading to the disproportional use of some resources over others.

In many cases, it is difficult at best to identify which habitat resources are truly available to any given individual over a given time period. Accounting for what is available, however, is an important part of documenting habitat selection and preference, because habitat preference is determined from a comparison of what is used with what is available. Habitat preference can be expressed in absolute terms (i.e., resource use exceeds, equals, or is less than availability), or preference can be expressed in relative terms (i.e., ranking of relative degree of preference when compared with the other available habitats). Also, habitat cannot only be described in terms of its quantity or relative abundance (usually in terms of area), but also in terms of its quality. Habitat quality is a somewhat elusive concept because it can be defined in numerous ways. From a ruffed grouse perspective, we refer to habitat quality as the ability of the habitat to provide the conditions capable of supporting successful reproduction and/or survival.

Life History Requirements

Ruffed grouse life history strategies are centered on the goal of maximizing reproduction and survival for several different life stages. Each stage may have its own specific habitat needs. In most cases, these habitat requirements differ between males and females because female habitat selection is often linked to maximizing foraging opportunities to maximize condition for reproduction. Males, in contrast, may select more for dense cover for security from predation (Whitaker et al. 2006). At times, differences in habitat selection may also exist between juveniles and adults. Starting in the spring (March–April), mate selection is centered on the drumming site. Drumming habitat availability determines largely where males select drumming sites. An adaptive strategy for nesting females is to select nesting sites away from drumming sites to avoid detection by predators. Apparent random use of mature forest may be a function of nesting in areas where there is a lot of habitat, thus making nest searching by predators more difficult.

After a successful nesting effort in May, the hen moves the newly hatched brood into adjacent areas that meet the brood's needs in terms of cover from inclement weather and predators, and provides invertebrates for food (see brood habitat descriptions in Chapter 4). The brood breaks up in early fall (September) as juvenile grouse begin to disperse. Dispersing juveniles can be found in a wide range of cover types as they move across the landscape and finally establish their permanent home ranges. Finally, in late fall, grouse move into their fall–winter habitats. In years of good mast crops, this may include use of oak-dominated stands. Later in winter, if acorn crops are depleted, grouse seek out a variety of leafy foods (evergreen leaves and ferns) and soft mast resources to sustain them through winter. In years of abundant hard mast (oak acorns and/or beech nuts), grouse may still use these resources until spring green-up (Long 2007). Unlike in regions further north, grouse in the Appalachians did not show a strong affinity for use of evergreen-dominated forests for thermal cover in winter for diurnal or nocturnal use (Whitaker 2003). We speculate that the thermal benefits in general were too limited to influence habitat selection patterns. We did observe daily movement to upper slopes prior to roosting in winter. We suspect such movements were tied to energy conservation accrued in slightly warmer upper-slope microclimates compared to cold-air-accumulating lower slopes (Whitaker and Stauffer 2003).

Methods of Habitat Analysis

Tracking radio-tagged grouse throughout the annual cycle served as the basis for documenting grouse habitat use in the Appalachian Mountains. Although the locations of radio-tagged grouse are invaluable for documenting habitat use at various scales, there are some limitations that need to be recognized. As discussed previously, grouse locations are determined usually by means of triangulation and consequently involve some degree of locational error. That error becomes significant when we attempt to attribute a given cover type to that location. For example, if the area of error associated with a given location covers 2 hectares (5 acres) in extent, that area may actually be composed of more than one cover type. This is especially true for a species like ruffed grouse that often use areas where there is a high degree of habitat interspersion. As a result, we may not be able to determine exactly which patch of cover the bird was in.

In some instances, grouse habitat can be described without the need to factor in positional errors associated with triangulated radio-tagged grouse. For example, in the case of nesting habitat, the actual nest itself is located so that nest-site habitat can be described in the vicinity around the actual nest without error. This is also true for habitat associated with drumming sites, and in some cases brood habitat when the telemetry gear is used to actually home in and get a visual sighting of the hen and brood.

Habitat Analyses. Three basic approaches have been used to document ruffed grouse habitat use and habitat selection in the Appalachians. We used basic descriptive statistics to describe grouse habitat use based on habitat parameters that were measured in the field (e.g., sapling stems per hectare around grouse nests). In addition to simple descriptive statistics, however, more advanced analyses entail use of statistical tools that allow for comparing habitats used with those that were available. For example, a method called compositional analysis (Aebischer et al. 1993) compares the composition of cover types associated with a grouse location with the total composition of the study area. If grouse select specific cover types to associate with, then the cover types close to grouse telemetry locations may differ from the overall composition of cover types that are available on a given study area.

Landscape-Level Habitat Selection

Habitat Availability. The study sites included in the ACGRP contained a great latitudinal range of habitat conditions in the Appalachian Mountains (from North Carolina to Pennsylvania) though the range of forest types studied was limited to those that grouse regularly occupied. Generally, this was limited to three major forest types: mixed-mesophytic forests, northern hardwoods forests, and more xeric (drier) oak-hickory forests. Although grouse will occasionally use other forest types (e.g., yellow pine stands), the majority of their use is restricted to the above types. Given the history of logging in the Appalachians, nearly all of the forests available to grouse had been previously logged sometime before or shortly after the turn of the twentieth century. In general, most forests on the study sites were middle-aged (40–80 years since logging), with a limited number of young age classes available (generally ~10% of stands were less than 15 years old) and a limited number of older age classes available (generally less than 10% of available stands were over 100 years old) (Table 9.1). Non-forested openings on the study sites were limited to less than 1% by area in most cases. Gated forest roads represented a significant habitat feature used by grouse across all sites, even though the relative area comprised in logging road habitat was limited to less than 5% of the total study area (Table 9.1). Habitat selection by grouse on the AGCRP study was a reflection of what was available on the study sites.

Oak-hickory sites in general were dominated by chestnut, white, red, scarlet, and black oaks, shagbark, pignut, mockernut, and bitternut hickories, white, pitch, Virginia, and tablemountain pine, eastern hemlock, red and sugar maple, and beech. Many of these forests were probably also dominated by American chestnut prior to introduction of the chestnut blight during the early 1900s. Mixed-mesophytic sites were co-dominated by red and sugar maple, basswood, sweet and yellow birch, black cherry, white ash, white pine, American beech, northern red oak, eastern hemlock, yellow buckeye, and yellow-poplar, depending upon individual site conditions. Northern hardwood forests, generally restricted to higher elevations in the southern Appalachians, were

Table 9.1. Distribution and age classes of forest types available on three Appalachian study sites in Virginia and West Virginia 1998–2001 (Endrulat 2003).

		Study Sites					
		VA2		WV1		WV2	
Age Class	Forest Type	Area (ha)	%	Area (ha)	%	Area (ha)	%
0-4	Hardwood	129	4	251	7	7	<1
	Softwood	41	1	0	0	0	0
	Other	1	<1	13	<1	2	<1
5-15	Hardwood	344	10	135	4	255	13
	Softwood	105	3	0	0	0	0
	Other	56	2	0	0	0	0
16-26	Hardwood	539	15	0	0	454	22
	Softwood	0	0	0	0	0	0
	Other	0	0	39	1	0	0
27-90	Upland Hardwood	1984	56	228	4	1181	58
	Cove Hardwood	0	0	215	4	15	1
	Mountain Hardwood	87	3	2489	57	0	0
	Softwood	6	<1	0	0	3	<1
	Mixed	105	3	0	0	2	<1
	Other	0	0	13	<1	6	<1
Roads		123	4	149	4	101	5
Total		3520	100	3532	100	2026	100

dominated by sugar maple, American beech, yellow birch, and black cherry.

In addition to forest type and age class, the spatial configuration of habitat patches is of primary importance to patterns of habitat use exhibited by ruffed grouse. A cover type patch can be defined as a distinct, recognizable area of vegetation that has similar structure and plant species composition throughout. Patches typically are characterized in a given area by different age classes of forest stands or different forest types, and can often be delineated by linear features on a landscape, including roads, streams, and other corridors. Several spatial characteristics of patches define grouse habitat suitability at a landscape scale. *Juxtaposition* refers to spatial arrangement of various cover types. That is, which cover types are adjacent to each other. *Interspersion* refers to the degree to which cover types are intermingled across the management area or landscape. Landscape linkages (corridors) are also important in defining ruffed grouse habitat suitability because linkages facilitate movement between habitat patches on either a diurnal or seasonal basis. Corridors important for ruffed grouse most often include riparian zones, sequential patches of successional forest, and woods roads.

Habitat Use. Ruffed grouse in the Appalachian Mountains can be found in virtually any forest type available, but their greatest densities occur in northern hardwood, mixed-mesophytic, and oak-hickory forests. A difference in how grouse used sites was seen between mesic and xeric (oak) forest types. Grouse, especially females, in oak-hickory forests made greater use of roads and bottomlands than grouse in mixed-mesophytic and northern hardwood forests. Grouse in mixed-mesophytic forests made greater use of clearcuts than in oak-hickory forests (Whitaker et al. 2006). We suspect grouse in oak-hickory forests are more nutritionally stressed than grouse in mesic forest types and, as a result, seek out higher-quality foraging opportunities along roadsides and in riparian zones to supplement the food resources found

What's up with oaks?

It is clear that the ecology of ruffed grouse in the Appalachians is influenced by oaks, and we thus need to understand something of this group of trees. As a group, oaks are arguably one of the most ecologically important trees in the continental United States. At least 58 different species of oaks occur within the United States, and their range includes most of the country, with the exception of the northern Rockies. In the eastern United States, and especially in the central and southern Appalachian region, oaks are the most common trees found in many forest stands. Portions of the central and southern Appalachian region contain a tremendous diversity of oak species (15 or more), some of the greatest concentrations of different species found throughout the oaks' range.

The oaks can be divided into two broad groups: the red oak group and white oak group. These divisions are based on differences in the flowering and acorn production characteristics between these two groups. The white oak group produces mature acorns in the fall from flowers that were present on the tree during the spring of that same year. Species belonging to the red oak group produce acorns from flowers that bloomed during the spring of the previous year. Common species within the red oak group include northern red oak, black oak, and scarlet oak. Other less common species include southern red oak, pin oak, and blackjack oak. The species in the white oak group most common to the central and southern Appalachians include white oak and chestnut oak. Others in the white oak group include the swamp white oak, chinkapin oak, post oak, and burr oak. In the eastern United States, red oak species tend to be slightly more abundant than those within the white oak group.

In the Appalachian region, oaks within both species groups tend to be most prevalent on drier, upland sites, such as hillsides and mountaintops having a southern, western, or southwestern exposure. Oaks tend to dominate these sites, comprising most or all of the trees present within the stand. However, they also can be a component of forest stands having a northern or eastern exposure, as well as stands found in the valleys and deep stream ravines common throughout the Ridge and Valley regions of the central and southern Appalachians. In these areas, oak species generally are evenly mixed with other tree species or are minor components of the overall forest stand.

One of the most ecologically interesting and valuable aspects of oaks is acorn production. Acorns serve as a critical food source for over 100 different species of wildlife, from white-tailed deer and black bear to chipmunks and deer mice. Even some songbird species such as blue jays eat acorns. Acorns within both species groups have a high-energy content and are easily digested. Those within the white oak group have a greater palatability than those of the red oak group, presumably because of the higher tannin levels present in red oak acorns. Tannins, which belong to a group of chemical compounds called phenols, give red oak acorns a bitter taste and they can interfere with various digestive and physiological functions in animals. Because most upland game birds lack a sense of smell, they typically select acorns based on their size rather than their taste or tannin content.

Acorn production varies considerably from tree to tree within the same species (for example, two white oaks or two red oaks). Some trees periodically produce very large crops in some years and have few or no acorns in other years. Other trees are consistently poor producers and have few if any acorns every year, even when other trees of the same species in the same stand have a large crop. Some trees appear to be genetically predisposed to having good acorn crops in years having favorable conditions for production while others are predisposed to producing few if any acorns regardless of other factors.

Acorn production across a forest stand for both the red and white oak species groups also varies considerably from year to year. Some years are "boom" years with large acorn crops within one or both species groups. Other years are "bust" years with little or no acorn production in one or both groups, and some years fall in between. The boom and bust crop years can occur in a somewhat cyclic pattern that is referred to as masting. Within the white oak group, good acorn crop years occur roughly every two to three years, and in the red oak group they occur about every five to six years. However, because some acorn production occurs almost every year and many factors influence acorn production within a forest stand, these boom and bust cycles are often difficult to detect. Several climatic factors, including spring temperatures when flowers are present, summer rainfall during the growing season, and frost events influence annual acorn production across forest stands and contribute to the annual variability in production patterns.

The pattern of acorn production within individual species groups in a given year is very similar across a large geographic area. Studies have shown that acorn production patterns from year to year are similar in forest stands separated by 300 miles or more. This broad geographic similarity in acorn production is referred to as ***synchrony*** and has been documented within both the red oak and white oak species groups. However, because of the differences in the flowering and acorn production characteristics that exist between these two groups, they rarely follow the same synchronous patterns.

The gypsy moth can be considered the most damaging agent to oaks in the Appalachian region. Gypsy moths are native to Europe and Asia, and were introduced into the United States in Massachusetts in 1868. Their range has continued to expand, with their southern extent reaching West Virginia, Virginia, and North Carolina. Gypsy moth caterpillars repeatedly defoliate trees and have killed oaks, especially northern red oak, in a wide area in the northeastern United States. Oak trees can recover from a single defoliation, but may be weakened enough to make them more susceptible to damage and death from other diseases or insects. Acorn production also will fail during years of moderate to heavy defoliation.

If left unmanaged, mature oak stands will gradually transition to stands dominated by other tree species, including maples, American beech, and black gum. Seedlings and saplings of these species tolerate the shaded understory beneath a stand of mature oak trees and can persist in the forest understory for long periods. Most oak species are considered either shade intolerant or intermediately shade tolerant, thus they do not survive as well in the shade as shade tolerant species. Maintaining oaks in a forest system requires active management, and some means to manage for oaks are described in Chapter 13.

—TODD M. FEARER

on the drier ridges and slopes. Grouse in mixed-mesophytic forests, in contrast, can afford to seek out habitats offering more cover (i.e., clearcuts) because of the lower levels of nutritional stress.

The age of forest stands and their juxtaposition, interspersion, and linkages are important predictors of relative use. Young forests (6–20 years old) provided the cornerstone of grouse habitat use on AGCRP study sites. Such early successional forests were selected in almost all cases, with stronger selection being shown by males and occurring when availability of young forests was rare on the landscape (Whitaker et al. 2006). In North Carolina, male and female grouse selected forest stands that were young (6–20 years old) and generally located on transitional sites close to upper-slope dry sites and lower-slope moist sites. The spatial configuration and size of cover-type patches appears to be important in defining habitat suitability for Appalachian ruffed grouse. Grouse in Virginia selected home ranges that had forest patch sizes smaller than the average for the study areas and in areas with greater than average amounts of edge between different patches of cover (Fearer and Stauffer 2004). Grouse in Virginia also selected for areas with greater than average cover type diversity. Grouse in North Carolina had smaller home ranges (i.e., they found all of their habitat requirements over a smaller area) in watersheds that had greater juxtaposition of selected cover types (midslope transition stands that were 6–20 years old). North Carolina grouse also had smaller home ranges when gated forest roads intersected these preferred habitats. These results have obvious implications for how managers and landowners might approach managing their forests for ruffed grouse.

Microhabitat Selection

Drumming logs. Male grouse select drumming sites usually adjacent to and above logging roads that traversed the study sites. In North Carolina, the logs actually selected for drumming were not unique in size and location when compared to other logs available on the study area (Schumacher et al. 2001; Table 9.2). Drumming sites were characterized by a display platform 30–45 centimeters (14–16 in) in height. These conditions were most often met on downed trees; however, other platforms such as rocks may be used on occasion. Drumming sites were located in areas with greater midstory densities and vertical cover than logs that were available randomly in the area. We found 85% of the drumming logs to be on or near a ridge top in mature (more than 40 years old) forest stands and significantly closer to gated forest roads than logs located at random in the forest. In spite of the fact that most drumming logs were in mature stands, male grouse made significant use of clearcuts during the breeding season (Whitaker et al. 2006). Thus juxtaposition of mature stands with clearcuts is important for male breeding season habitat selection. The drumming site is the focal point in a male's home range, and is used in both spring and fall. In most cases, however, the availability of potential drumming logs themselves did not limit grouse distributions on ACGRP study sites. Drumming sites, however, were selected in specific locations on the landscape, presumably because they were positioned to maximize potential contact with females during the breeding season while minimizing potential for predation.

Table 9.2. Average habitat characteristics for drumming logs and random logs located on Wine Spring Creek Ecosystem Management Area, North Carolina, 1999–2001 (Schumacher et al. 2001). Vertical vegetation density and mid-story density were significantly different between drumming logs and random logs. From these data it is clear that males are selecting logs that have greater density of vertical vegetation and mid-story stems.

Habitat Variable	Drumming logs	Random logs
Height (cm)	50.0	51.9
Diameter (cm)	50.5	52.8
Length (m)	8.3	7.6
Slope (%)	26.9	28.3
Moss Cover (%)	34.8	44.6
Vertical Vegetation Density (%)	41.4	25.7
Basal Area (m^2/ha)	15.4	18.1
Understory Density (stems/ha)	12,433	10,302
Mid-story Density (stems/ha)	6,805	3,438

Nest Sites. Hens selected nest sites generally from whatever forest types were available on the area. Nest-site selection appears to be more related to structural habitat characteristics than forest species composition. Appalachian grouse nested in forest stands with greater basal areas and also in stands with significantly more coarse woody debris (logs and tree limbs on the ground) than average conditions on the study area (Tirpak et al. 2006; Table 9.3). Selection for these characteristics makes biological sense because hens place nests under or against logs, stumps, or living trees. In addition, nests tended to be closer to roads than other locations within the hen's home range; 55% of nests were located within 10 meters of a road compared to only 17% of systematic points within the hen home range were within 10 meters of a road (Table 9.3). Selection for roadside nesting sites may reflect the desire to use roads as corridors for foraging while the hen is incubating, and for moving the brood from the nest site to high quality brood habitat.

Table 9.3. Habitat characteristics at nest and systematically located sites, 1995–2002 (Tirpak et al. 2006). The data are based on 73 nest sites and paired samples located on 5 study areas.

	Mean Values	
Habitat Variable	Nest Site	Systematic Site
Basal area (m^2/ha)	24.7	19.7
Deciduous canopy cover (%) of trees (≥8 cm dbh)	80.8	83.5
Coniferous canopy cover (%) of trees (≥8 cm dbh)	5.1	6.5
Ground cover (% herbaceous or woody vegetation ≤1m tall)	47.2	49.8
Coarse woody debris cover (% dead woody vegetation ≥15 cm diameter)	17.3	9.2
Stems (<8 cm dbh and ≥1.5 m high)/ha	13,382	10,701
Distance to road or opening class (%)		
Close (<10 m)	54.8[a]	17.2
Moderate (11–100 m)	20.5	74.5
Far (>100 m)	24.7	8.3
Timber size class (%)		
Sapling (<12.5 cm dbh)	34.2	38.2
Pole (12.5–27.8 cm dbh)	37.0	35.6
Sawtimber (>27.8 cm dbh)	28.8	26.3
Midstory (>1 m high to bottom of canopy) volume class (%)		
Open (<20%)	28.8	3.5
Moderate (20–50%),	67.1	90.0
Closed (>50%)	4.1	6.5
Understory (≤1 m high) volume class (%)		
Open (<20% woody and <30% herbaceous)	49.3	43.5
Herb (>30% herbaceous vegetation)	31.5	37.5
Wood (>20% woody vegetation)	19.2	19.0
Slope class (%)		
Gentle (0–10%)	43.8	49.1
Moderate (11–30%)	47.9	44.8
Steep (>30%)	8.2	6.1

[a] represents the percent of total within each class of variable.

Brood Habitat. The brood period is critical for grouse population recruitment and brood survival is a key factor limiting population growth, therefore habitat use during this period is very important. Brood habitat is influenced by the needs of the grouse chicks during late May–August to locate invertebrates and succulent, green vegetation for food while avoiding potential predators. To meet these needs, hens in the Appalachians move broods into areas with well developed herbaceous understories that support relatively high invertebrate densities under well-developed overstory canopies and modest understory woody stem densities (Table 9.4). This pattern is different from studies conducted within the range of aspen in the Lake States, where brood habitat is often characterized as areas with high sapling stem densities (Gullion 1977, Kubisiak 1978). In the Appalachians, there are areas in the forest with greater stem densities than those selected by hens as brood habitat. However, it is the balance between cover (moderate stem densities and herbaceous cover) and food (invertebrates) that hens appear to seek. These conditions, as a result, are somewhat unique in typical Appalachian forests. Broods frequently move great distances from nest sites to reach these preferred conditions for the brooding period. Often these movements are tied to the presence of forest roads and other natural corridors traversing the forest. These movements often have survival consequences, as broods that don't travel as far tend to experience greater survival rates (Tirpak et al. 2005). Thus, good interspersion and juxtaposition of nesting habitat with brood habitat has positive survival benefits for grouse.

Table 9.4. Brood habitat characteristics in North Carolina (Fettinger 2002) and Virginia–West Virginia (Haulton et al. 2003) compared with random points located within 100 m of brood locations 1997–2001.

	North Carolina		Virginia–West Virginia	
Habitat Variable	Brood Mean	Random Mean	Brood Mean	Random Mean
Basal Area (m^2/ha)	20.7	20.9	15.9	17.7
Canopy Cover (%)			82.0	81.0
Saplings (stems/ha)[a]	5975	4598	3822	3581
Ground Cover (%)[b]	54.0	37.0	62.0	49.0
Vertical Cover (%)	55.0	44.0		
Distance to Road (m)	130.0	111.0		
Distance to Opening (m)	444.0	446.0		
Distance to Cut (m)	56.0	58.0		

[a] Saplings defined as < 11.4 cm dbh in NC and <8 cm dbh in VA-WV.
[b] Ground cover included only herbaceous plants in NC, and included herbaceous and woody in VA-WV.

Fall dispersal. Fall represents a critical time period during which broods break up and juvenile grouse disperse. Habitat use at this time can be described based on what juveniles are selecting as opposed to what adult grouse with established home ranges have previously selected. In North Carolina, juvenile and adult females selected similar vegetation types during the fall (Jones 2005). Although juvenile females were dispersing, they still showed similar habitat selection patterns as their adult counterparts. In contrast, juvenile males showed substantial differences in habitat selection from adult males. Adult male habitat selection during fall in North Carolina was similar to what was observed throughout the annual cycle, selecting for young (6–20 year old) stands, usually associated with roads (Jones 2005). Juvenile males, in contrast, showed little tendency to select for particular habitats at all; juvenile male habitat use did not differ from what was available on the study area in general. Based on these results, a picture emerges of adult males staying in their home ranges associated with their drumming sites, perhaps to maintain their territorial dominance of that area, as juvenile males are seeking potential sites of their own (Whitaker 2003). Juvenile males, excluded from already occupied adult male territories, disperse across the landscape, seeking out areas with vacant territories to settle. All grouse during this period show preference for use of gated forest roads. Even in the case of juvenile males, roads may serve an important function as dispersal corridors. By December, the fall shuffle is complete and most juvenile males have settled into their permanent home ranges.

The availability of hard-mast crops influences fall habitat selection by ruffed grouse. Selection of forest roads, particularly by female grouse, was stronger during fall–winter if mast crops were poor (Whitaker et al. 2006). Roads may provide important alternate foraging habitat for grouse during fall, especially when hard mast is limited.

Winter. Grouse habitat requirements during winter are associated with meeting nutritional needs while avoiding predation. In North Carolina, habitat use did not differ by age class; juvenile grouse settled into home ranges that were very similar to those of adults. Males consistently continued to select young xeric stands (6–20 years old) associated with roads (Jones 2005). Selection in winter for high stem densities undoubtedly is related to the need for cover for predator avoidance. Elsewhere in the Appalachians, male grouse selected mesic bottomlands during the non-breeding season, including winter. This pattern of habitat selection may have been influenced by increased foraging opportunities in this cover type in winter, especially in poor hard mast years (Whitaker et al. 2006). In contrast, females used a broader cross range of age classes of stands, but still were generally associated with xeric stands and roads. Because females are not tied to a particular site or territory during this time of year, they have the ability to use a larger home range that encompasses a greater range of habitats, in association with pursuing optimal foraging strategies. Female habitat selection in winter must balance the need for maximizing nutritional gains in preparation for the upcoming breeding season with the need to avoid predators.

In summary, throughout the year, ruffed grouse reflect their classification as a "forest grouse." They require a diversity of forest stand ages to meet their needs on an annual basis; but, it is clear that the highest quality grouse habitat will have a relatively large proportion of young (6–20 years) forest present, along with some high-quality herbaceous cover to support broods and more mature stands for breeding. When we consider particular locations within the stands, our detailed microhabitat measurements confirm that denser is better. Reducing all the arcane habitat measurements and statistical analysis to something practical, we get: The more difficult it is to walk through, the better the habitat is.

10

Home Ranges and Space Use

Darroch M. Whitaker, John M. Tirpak, William M. Giuliano, and Dean F. Stauffer

How much space does a grouse need to breed or overwinter safely? Is this a fixed quantity or is it influenced by such factors as food availability or risk of mortality from predators or hunters? How far does a grouse settle from its natal range where it hatched? What factors affect the distance a grouse travels in a day, month, or year? How much territory does a male grouse defend around its drumming log? Does the amount of movement affect survival or reproductive success of a grouse?

These are all critical questions in the ecology and life history of grouse, and are of direct importance to grouse management and conservation. For example, it has been shown that increases in movement rate and home-range size are related to reduced survival for ruffed grouse (Thompson and Fritzell 1989, Clark 2000). Further, an understanding of factors that affect movements and home ranges can provide insight into a species' habitat ecology and social systems, and point to factors limiting the grouse population in a region. Consequently, information of this nature can give effective guidance to management decisions. In this chapter we review a range of topics relating to the spatial ecology of Appalachian ruffed grouse, drawing on numerous studies conducted through the ACGRP, as well as on other information available in the scientific literature.

The general topic of space use overlaps with information presented in other chapters of this book. Space use is inseparable from many aspects of a species' habitat ecology, such as the use of seasonal habitats and the distribution of resources such as food and roosting sites. Similarly, dispersal is a key aspect of a species' spatial ecology; given that dispersal is the focus of the next chapter we give it only passing coverage here.

Terminology and Key Concepts

A home range of an animal has been defined as the area traversed during its normal activities for a defined period of time (Kenward 2001, Kerhonan et al. 2001). The time period during which the home range is used is an important part of this definition, as an animal may shift its home range over time, for example between the breeding and nonbreeding period. The perimeter of an individual's home range may be only generally perceived by the animal itself, and rarely acts as a fixed boundary on its movements (Powell 2000). Regular exploratory forays or sallies outside of the home range can be common, and animals will often readily exploit new areas as their needs, environmental conditions, or resource availability change. A territory is a portion of an individual's home range that is defended from other members of the species, most often those of the same sex (Powell 2000). A territory can encompass the entire home range, but typically represents only a limited portion of it.

As mentioned, animals may only generally perceive home-range boundaries, thus decisions by researchers can influence the perceived home-range size estimated from a set of animal locations. Consequently, estimates of home-range area can be imprecise and comparisons of home-range size between different studies can be problematic (Lawson and Rodgers 1997, Powell 2000). A variety of quantitative techniques have been developed to estimate home-range boundaries from sets of animal location data, and of these, an approach known as the fixed kernel method appears to be the most accurate and robust (Worton 1989, Seaman et al. 1999, Powell 2000). This approach allows the researcher to develop contours based on the animal's locations that represent a specified proportion of their home-range use (Figure 10.1). An assumption in home-range estimation is that the animal has settled and is restricting its movements within a particular area. Dispersal by an animal is a one-way directional movement through an area that is not revisited, and areas traversed during dispersal are not considered part of an animal's home range. Thus, efforts must be made to exclude dispersal movements when identifying home ranges, as failure to do so can lead to exaggerated and erroneous home-range estimates.

Movement patterns exert a strong influence on a species' spatial ecology. Natal dispersal is movement from an individual's natal home range to the area where it first attempts to breed, and is undertaken by most juvenile grouse. Breeding dispersal, where an adult relocates to a new breeding site, is less common and can result from low habitat quality or poor reproductive success in previous breeding seasons.

Site fidelity, also known as philopatry, is the

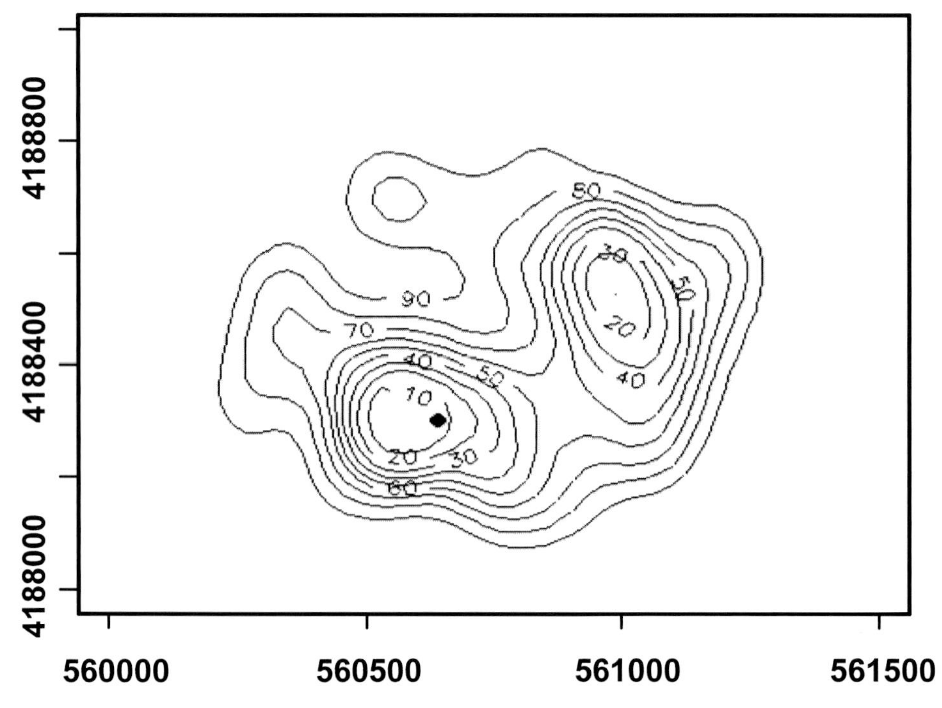

Figure 10.1

Home range contours estimated by applying the fixed kernel technique to a set of 62 locations from a female ruffed grouse on the WV2 site, which was followed during summer 1999. The black point indicates the location of its nest, from which three of 11 eggs hatched and one chick survived to five weeks of age. Contour lines delineate the boundaries of the smallest area accounting for the specified percentage of total use. Numbers on the X and Y axes are UTM (Universal Transverse Mercator) coordinates, which are measured in meters.

tendency of an animal to remain in a given area over time. However, site fidelity should not be viewed simply as the opposite or absence of dispersal. Rather, site fidelity is the choice by an animal to remain in a given area, and is more appropriately contrasted with nomadism or vagrancy, where an animal does not settle in any one area for an extended period. Fidelity to a particular site is shown to some degree by most animal species, and may have many advantages. These might include reduced time spent searching for resources such as food or mates, improved knowledge of escape cover, and dominance in territorial disputes. With reference to ruffed grouse, for example, Yoder (2004) reported that site familiarity led to a large reduction in predation risk.

The ACGRP Database

The general research methods and data collection protocols followed by ACGRP co-operators have been described in Chapter 2. Information presented in this chapter was derived primarily from ACGRP radio-tracking activities. This was one of the largest animal location datasets ever collected and allowed researchers to address many questions that had previously been difficult to address. Analyses of ACGRP radio-tracking data were first presented in Whitaker (2003), and readers may consult that source for a detailed description of methods of data collection, error testing, data processing, and analysis. Briefly, from fall 1996 through fall 2001 ACGRP personnel attempted to triangulate each radio-equipped grouse at least twice weekly. Locations were estimated from sets of telemetry azimuths using Lenth's approach (Lenth 1981). Results from a study using transmitters having known locations (White and Garrott 1990) indicated that mean location error was less than 76 meters (Whitaker 2003). Virtually all locations were collected between dawn and dusk, and thus reflect daytime activity.

For analyses involving home ranges, we divided the year into fall–winter (September 1–March 31) and spring–summer (April 1–August 31) seasons to approximate the nonbreeding and breeding seasons, respectively. Grouse were considered juveniles up until September 1 of their second fall (≈ 15 months of age), and were considered adults thereafter. In cases where we wanted to distinguish between the summer in which a juvenile grouse hatched and the subsequent summer when it first had the opportunity to breed, we refer to these as Hatch Year (HY) and After Hatch Year (AHY), respectively. Home-range boundaries were estimated using the fixed kernel method (Worton 1989, Seaman and Powell 1996). A minimum of 30 locations was used to estimate home ranges. Because definitions of home ranges assume that an animal has settled in an area, and because inclusion of locations collected outside the home range during dispersal dramatically inflate home-range area estimates, we removed presumed dispersal movements prior to analyses.

Finally, as presented previously, two forest associations were prevalent across the region encompassed by the ACGRP. Oak-hickory forests dominated cover on the KY1, RI1, VA1, VA2 and WV2 study sites, and were characterized as being relatively dry and dominated by various species of oaks and hickories, as well as by rarity of cherry, birch, and aspen (Braun 1950, Whitaker 2003). Mixed-mesophytic forests dominated cover on the MD1, NC1, PA1, VA3, and WV1 sites, and were characterized by having moister soils and better-developed understory vegetation. While still supporting an abundance of hard-mast producing trees such as oaks and beech, mixed-mesophytic forests were distinguished by having a greater abundance of many northern hardwoods, particularly cherry, birch, and aspen. These trees are important grouse foods where available (Servello and Kirkpatrick 1987), and accounted for 22% of canopy trees on these sites (Braun 1950, Whitaker 2003).

Home Range and Territory Size of Appalachian Ruffed Grouse

As reported for other regions (Archibald 1975, Thompson 1987, Clark 2000), home ranges of juvenile and adult female grouse typically were two to three times larger (25–30 ha) than those of adult males (Figure 10.2). Also consistent with other studies (Archibald 1975, Maxson 1978, Thompson 1987), home ranges of all sex and age classes were larger during fall–winter than spring–summer (Figure 10.2). Juvenile males showed the largest difference in size of seasonal home ranges, occupying home ranges comparable in size to those of females during fall–winter but only slightly larger than those of adult males during spring–summer. On average, home ranges of all sex and age classes of grouse were slightly smaller on study sites having mixed-mesophytic forests than on sites having oak-hickory forests, though this difference was not statistically significant (Figure 10.2). There was considerable variation in home-range size across individuals and study sites (e.g., Figures 10.3 and 10.4), and approximately 10% of fall–winter home ranges exceeded 100 hectares.

Though methodological inconsistencies make comparisons of home-range size across studies problematic (Lawson and Rodgers 1997), authors have consistently reported that home ranges of ruffed grouse inhabiting the southern Appalachians are larger than those of grouse in northern forests (Godfrey 1975b, White and Dimmick 1979, Epperson 1988, Thompson and Fritzell 1989, Neher 1993, Scott et al. 1998, Clark 2000, Whitaker 2003). Reported differences are greatest for females, particularly those rearing broods. The relatively large home ranges used by grouse in the southern Appalachians, as well as typically lower population densities, have led biologists to suggest that quality of ruffed grouse habitat is generally inferior in this region. However, prior to the ACGRP there was limited information on factors that were associated with variation in the home-range size of ruffed grouse, so this view of habitat quality was largely speculative.

No data on the size of territories defended by male grouse were collected through the ACGRP. However, other studies have reported that territories, which are centered on display sites (drumming logs), comprised relatively small proportions of male home ranges, and averaged 2–3 hectares or 20–30% of the overall home range (Archibald 1975, Rusch et al. 2000). Further, drumming sites and hence territories of neighboring males typically were well spaced and averaged 255 meters apart in Pennsylvania (Lovallo et al. 2000). When population densities were high, approximately 50% of males used multiple drumming sites, and these additional sites were located within 100 meters of the primary site (Lovallo et al. 2000). Females do not defend territories, and there is considerable overlap in home ranges of neighboring grouse regardless of sex (Maxson 1978, Rusch et al. 2000).

Factors associated with variation in ruffed grouse home-range size

Home-range size is a fundamental aspect of an animal's habitat ecology, having important implications for energetics, survival, time budgets, movements, and spatial relations with other animals. Larger home ranges may be costly in terms of time and energy allocated to travel, while also potentially increasing encounter rates with predators and competitors. Consequently, we would expect that under most circumstances animals should attempt to use the smallest adequate home range. Further, home-range size should be positively related to resource needs, and inversely related to resource availability. However, there are also circumstances under which an individual might expand its home range beyond that sufficient to meet nutritional needs alone. These include, for example, finding mates, acquiring breeding territories or display sites, and displacement by territorial individuals of the same species. Consequently, knowledge of factors associated with variation in home-range size can help identify limiting resources and point to differences in habitat ecology and resource needs between populations.

Consider for a moment the numerous fall–winter home ranges of ruffed grouse monitored on the VA1 study site from 1997–2001 (Figure 10.4). Although this plot of home ranges is quite congested, when looking closely, we can see that grouse in some areas typically had relatively small home ranges, while those on other portions of the study site had consistently large ranges. In addition to this seemingly non-random pattern of variation in home-range size within study sites, there was considerable variation in home-range size among study sites (Figure 10.3). What underlying ecological factors could lead to these patterns in home-range size, and what are the implications

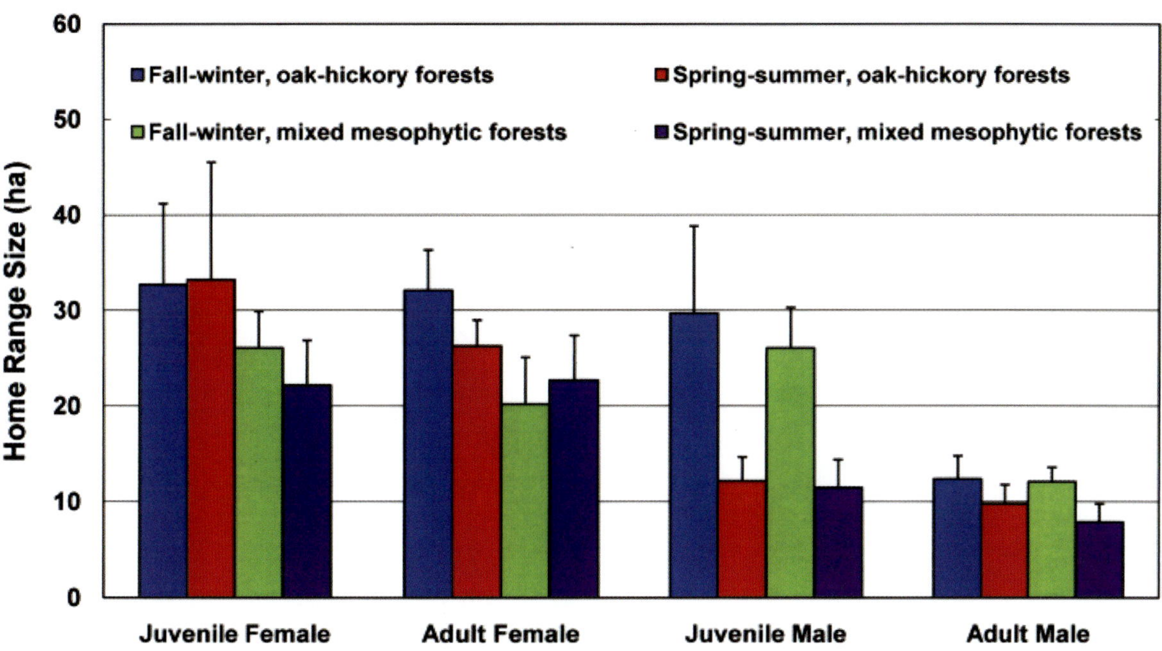

Figure 10.2

Mean home-range size of Appalachian ruffed grouse during fall-winter and spring-summer on 10 ACGRP study sites (1996–2001; data from Whitaker 2003). Means were estimated separately for sites typified by oak-hickory (n = 5) or mixed mesophytic forests (n = 5), and estimates are of the area of 75% fixed-kernel home ranges. Bars indicate one standard error.

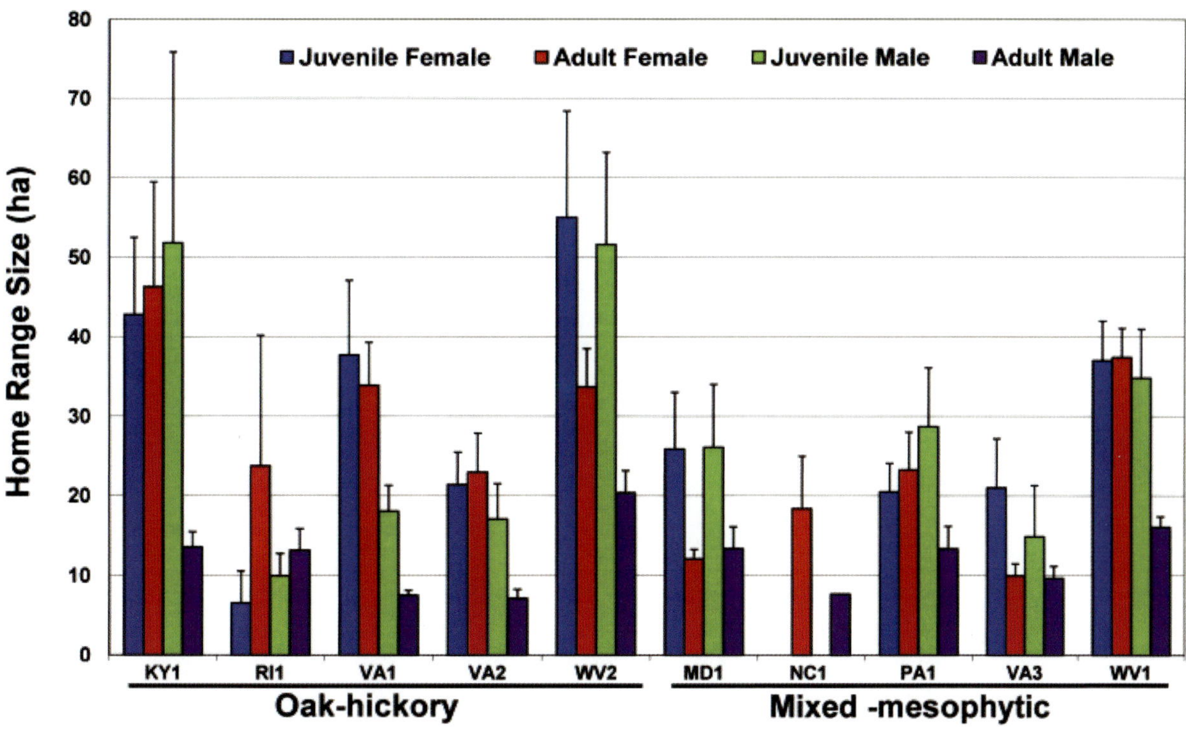

Figure 10.3

Mean size of 75% fixed-kernel home-range estimates for Appalachian ruffed grouse on 10 ACGRP study sites during fall-winter, 1996–2001 (from Whitaker 2003). Each study site was characterized by having either oak-hickory or mixed mesophytic forests. Bars indicate one standard error.

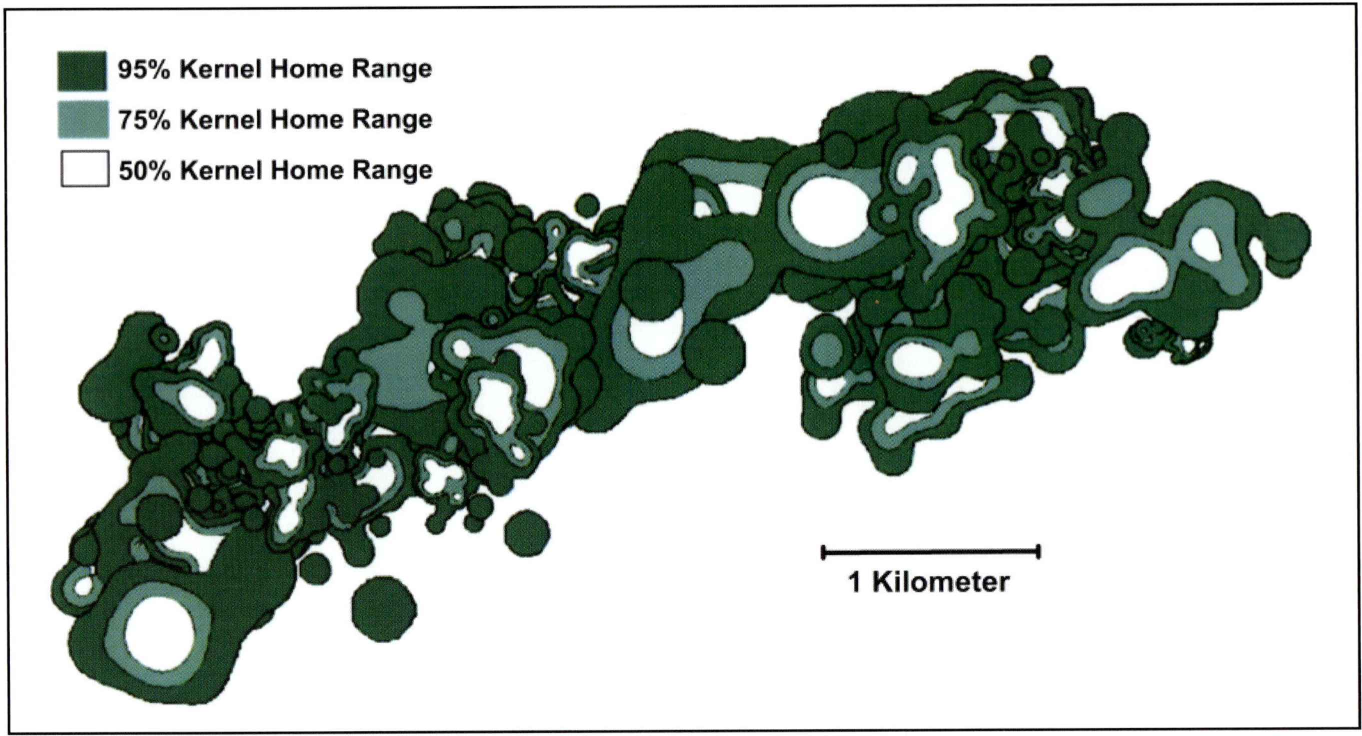

Figure 10.4

Figure 10.4. Fall-winter home ranges of all grouse monitored on the VA1 study site in George Washington National Forest, 1997–2001.

for the grouse inhabiting these various areas? An obvious explanation would be that habitat quality varies across space and through time, but to test this one must first consider the aspects of the environment that determine habitat quality.

We evaluated factors believed to affect variation in ruffed grouse home-range size across 10 ACGRP study sites (Whitaker 2003), and investigated the influence of landscape structure on home-range size on the VA3 study site (Fearer and Stauffer 2003). We employed two approaches in investigating patterns in home-range size. First, we used statistical models to evaluate factors influencing patterns across more than 1,000 home ranges collected across all study sites. Second, for those grouse that were followed during multiple seasons, comparisons between pairs of home ranges from individuals followed during more than one season were used to evaluate the influence of factors that differed for these individuals over time (age, season, breeding success, and food availability). Both approaches identified similar ecological patterns.

As mentioned previously, study site, season, sex, and age are important determinants of home-range size for ruffed grouse (Figures 10.2 and 10.3). A number of other variables also were associated with variation in home-range size, and these related to such diverse ecological factors as population density, reproductive status of females, forest type, habitat composition, food availability, and hunting pressure. For example, female grouse with broods had much larger home ranges than those that had experienced reproductive failure, presumably in response to the specialized habitat needs and increased food requirements of broods (Figure 10.5; Thompson et al. 1987, Scott et al. 1998, Fettinger 2002, Haulton et al. 2003). Only when hens experienced reproductive failure during summer were female home ranges comparable in size to those of adult males, suggesting that when females were relieved of parenting they attempted to minimize the extent of their movements (see also Maxson 1978).

Social factors can have important effects on home-range size, and one might expect these pressures to be influenced by territoriality. Although for the most part solitary, ruffed grouse do not occupy the environment independently of one another. During fall and winter when population density was relatively high (as indicated by high trapping success) mean home-range size for juvenile male grouse was over 25 hectares, compared to about 10 hectares when populations were low. No other sex or age class adjusted home-range size in response to population density. This suggests that juvenile males may compete with adults for drumming sites. This agrees with Gullion (1981b), who reported that the proportion of non-territorial males (primarily juveniles) in grouse populations was positively correlated with population size. Male grouse aggressively defend drumming logs, and established adults occupy a preferred subset of these display sites (Gullion and Marshall 1968, Rusch et al. 2000). As a result, when population densities are high, juvenile males likely have to range farther and monitor more occupied sites in hope of eventually obtaining a preferred drumming site (Marshall 1965, Archibald 1976, Gullion 1981b). Anecdotal behavioral observations that we were able to make also support the view that juvenile males continually monitor their neighbors and prospect for unoccupied high-quality drumming sites. On two occasions, established adult males we were tracking were killed during fall, and nearby radioed juvenile males abandoned their home ranges and relocated to the vacant display site within 48 hours.

When averaged across study sites and years, home-range size differed only slightly between

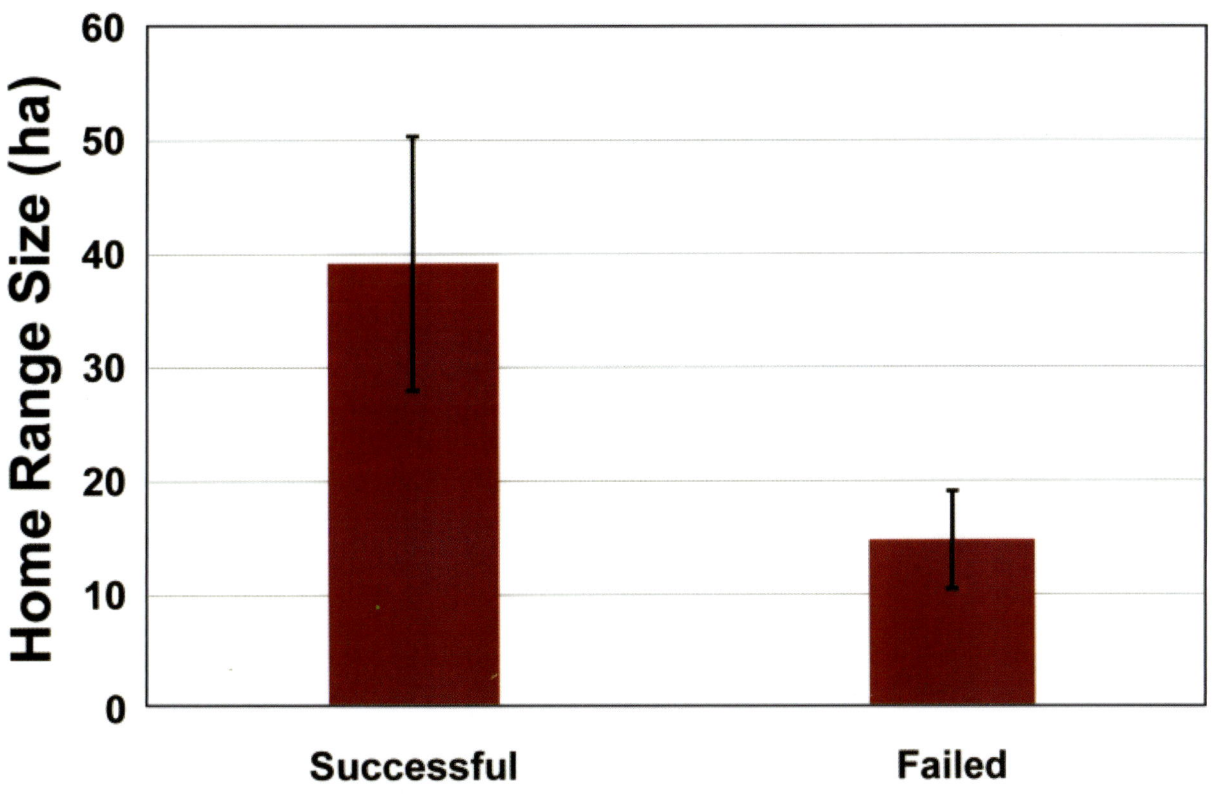

Figure 10.5

Comparison of size for pairs of home ranges from individual female ruffed grouse that successfully raised broods during one breeding season but experienced reproductive failure in another on ACGRP study sites, 1996–2001 (n = 12 individuals). Bars indicate one standard error.

Home Sweet Home Range —Dean F. Stauffer

We estimated the location of many of the grouse we radio tagged remotely, using radio-telemetry. One of the analyses we conducted with these locations was to estimate where the home range was for each individual. The concept of home range for wild animals has been around for a while; in the simplest sense, the home range is the area where an animal spends its time. Home ranges can be described for different periods of time such as a week, month, season, or year. Once the home-range location has been determined, we can estimate the area used, habitats used, and so on.

A number of different approaches to estimating the home range of an animal have been proposed over the past six or seven decades. Regardless of which approach is used, we first begin with a set of locations. For our data sets, the locations were determined by triangulation of three or more azimuths that were taken from the telemetry azimuths. The technique of estimating the location used the Lenth Estimator. Here is a representative set of locations for a female grouse on the VA1 study site:

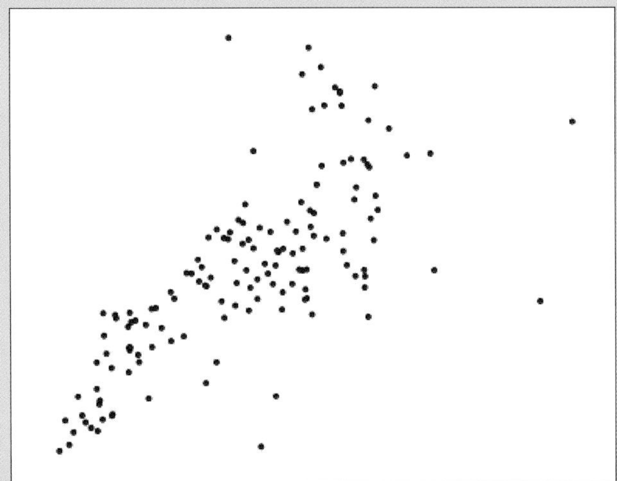

We can see that this set of points tends to be fairly well clumped, but several points might represent forays of the bird to locations outside of the main area she used. One of the older, traditional means of estimating a home range is called the Minimum Convex Polygon (MCP). The MCP simply connects the outside locations to define the home range. Here is the MCP home range for this set of points:

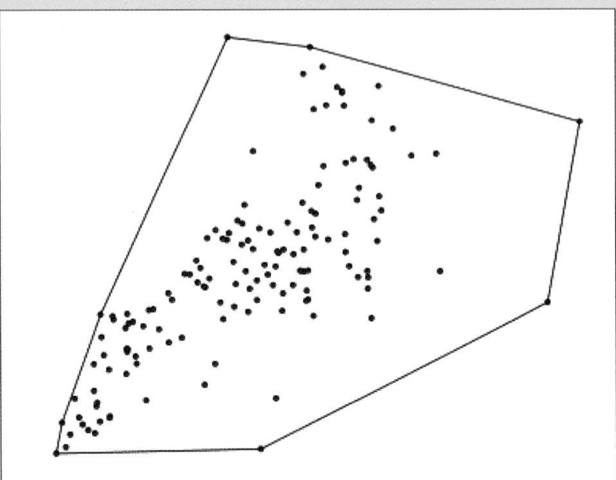

The total area within the MCP is about 143 ha (350 acres). It is clear, though, that there are portions of this MCP home range not apparently used much by the bird, and we have included "empty" space within our home-range estimate that the female may not have used frequently. This has been one of the primary criticisms of MCP as a home-range estimator over the years. Several other approaches were tried during the 1970s and 1980s; in the early 1990s an approach based on what are called kernel estimators became the most accepted method. The kernel estimator uses a probability density function to estimate the areas that are used most frequently. Several different means of applying kernel estimators exist, and researchers have not all come into agreement as to which variation is the "best." For our work, we used the fixed kernel estimator.

The kernel estimator basically analyzes the distribution of points, and develops a set of contours that reflect where the greatest "mass" of points exists. These approaches allow the researcher to develop home-range estimates for different levels. For example, applying the fixed kernel to our hen's locations, and requesting a 95% home range, results in this:

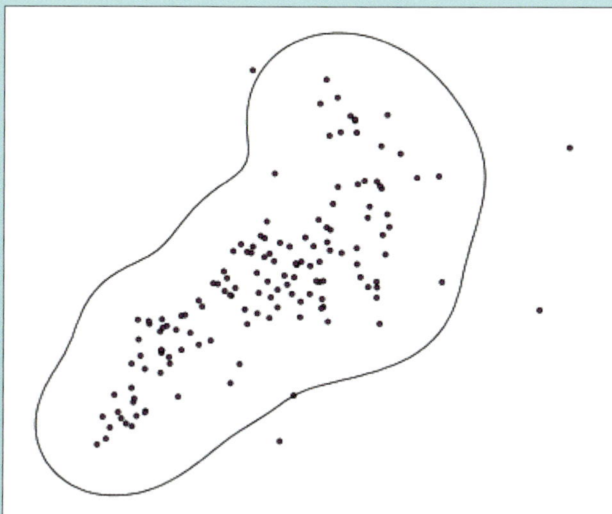

The area within this home range is about 157 ha (385 acres). We can see that some of the "outlier" points have been dropped from this home range, but because of the statistical approach used by the program, some "unused" area is also included, reflecting our uncertainty in the exact location of the points. Researchers often calculate different levels of home-range use. For most of our work, we used the 75% home range to reflect what we believed is a reasonable measure of the area primarily used by each bird. This home range would encompass 75% of the total volume of area, based on location density, used by the bird. Here is the 75% home range for our hen:

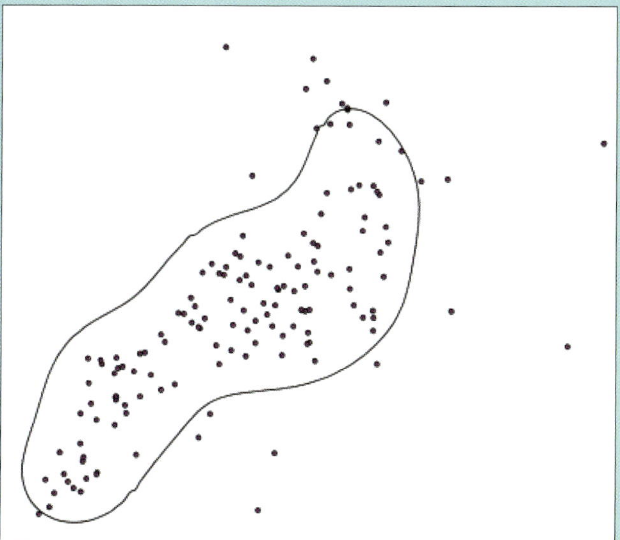

We see that a few more points were left out of this home range, yet most of the points are still included. The area of this home range is about 62 ha (152 acres). We sometimes wish to determine the "core" area of the home range — when we did that on the ACGRP, we calculated the 50% home range, representing the area that has 50% of the volume of the locations. The 50% fixed kernel contour for our Virginia hen is:

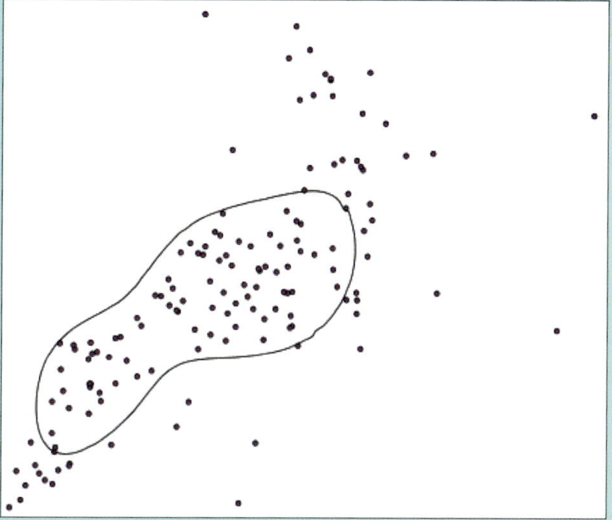

We see that quite a few more points have been eliminated, and the area of this contour is 31 ha (76 acres). This approximates the "core" of the home range, as the 31 ha is about 20% of the area of the 95% kernel home range, yet it contains 50% of the observations. Clearly, this reflects the core of where this bird spent its time. For easy comparison, here are the three kernel estimates plotted together:

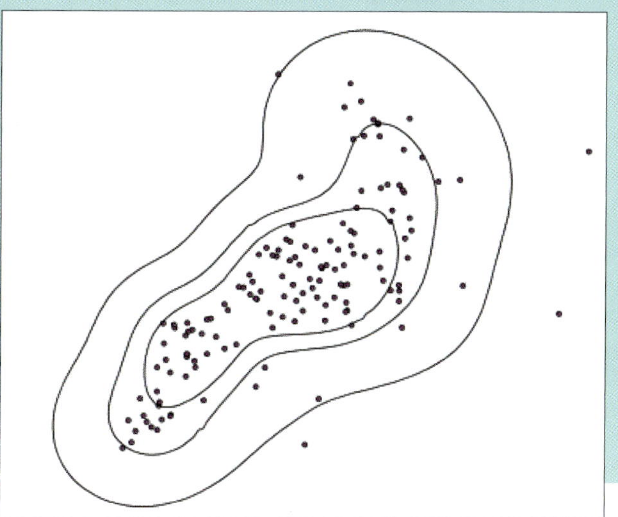

oak-hickory and mixed-mesophytic forests (Figure 10.2). However, there were important differences in the ecological factors that were associated with variation in home-range size between these forest types. Most notably, fall hard-mast crops (i.e., acorns) had a large influence on home-range size of adult ruffed grouse on sites having oak-hickory forests, but no measurable effect in mixed-mesophytic forests (Figure 10.6). While during years of abundant hard mast home ranges were similar in size in these two forest types, during poor mast years, adult grouse increased their home ranges over 2.5 times on sites having oak-hickory forests. This influence continued into the breeding season for females, even though most hard mast from the preceding fall had either been consumed or had germinated by late spring. This leads us to speculate that either this influence of mast crops occurs early in spring in the form of reduced prenesting movement, or that females are able to carry some benefit of good mast crops, such as high body-fat reserves, well into the breeding season. Other authors have suggested that Appalachian ruffed grouse are under strong nutritional constraint, and availability of sufficient hard-mast foods may be important for maintenance of body condition through winter and subsequent reproductive success (Chapter 7, Norman and Kirkpatrick 1984, Servello and Kirkpatrick 1988, Devers 2005). The association between hard-mast crops and home-range size provides strong evidence that hard mast is in fact a limiting resource in oak-hickory forests. Other ACGRP research supported this, showing, for example, that in oak-hickory forests fall hard-mast crops had a large effect on grouse reproductive success and habitat selection (Devers 2005, Whitaker et al. 2007). Increased home-range size due to poor hard-mast crops also suggests that availability of this resource could affect predation rates, though this effect was not observed (Devers 2005).

Given the strength of the relation between

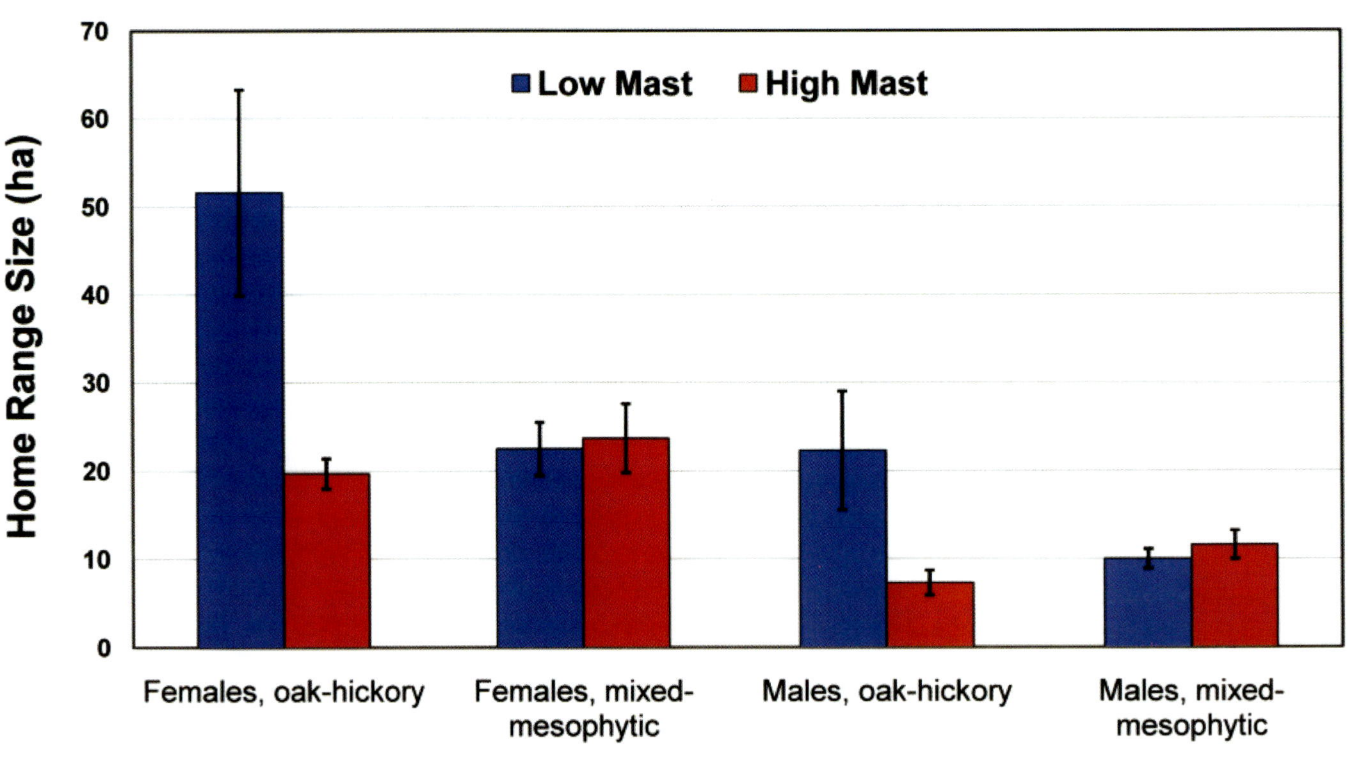

Figure 10.6

Figure 10.6. Size of fall-winter home ranges from individual adult ruffed grouse compared between years having high and low mast crops, 1996–2001. Study sites were grouped by dominant forest type, and 16–23 individual grouse were included in each comparison. Bars indicate one standard error.

mast crops and home-range size in oak-hickory forests, it is interesting that grouse in mixed-mesophytic sites were unresponsive to variation in mast production. In northern forests at the core of the species range, grouse feed heavily on buds of aspen, cherries, and birch during winter, and these provide a reliable and accessible source of high-quality food (Servello and Kirkpatrick 1987). This leads us to speculate that the 10-fold higher abundance of these trees, particularly birch and cherry, in mixed-mesophytic forest types (Whitaker 2003, Chapter 2) buffers any response by grouse to changing mast crops by providing a stable, accessible supply of high-quality food during winter (see also Servello and Kirkpatrick 1987). In contrast, alternate winter foods in oak-hickory forests consist largely of low-quality evergreen leaves.

Because it is assumed that under most conditions it is advantageous for animals to occupy the smallest adequate home range, it's reasonable to expect that an increase in availability of high-quality habitat should result in a reduction in home-range size because a smaller area will provide sufficient resources. In line with this reasoning, the three habitat features that showed the strongest relation with home-range size—clearcuts, access routes, and mesic bottomlands—were all selected by ruffed grouse on our study sites (Whitaker et al. 2006). It is important to note that these relations were with the proportion of a habitat feature found within an individual's home range, and that there was no relation between home-range size and the overall availability of a habitat in the landscape. Perhaps this should be expected, as animals do not settle at random across landscapes, but rather seek out localities affording favorable conditions. It appears that grouse adjusted their home-range size to suit their immediate surroundings, and not the average conditions across the landscape.

For all sex and age classes of grouse on all study sites, we observed an increase in proportional cover of clearcuts in smaller home ranges, as well as increased densities of access routes (roads and trails). It is well known that ruffed grouse preferentially select dense, early successional stands, and it is generally assumed that these serve as escape cover (Bump et al. 1947, Rusch et al. 2000). Recent research has found that survival was higher for grouse whose home ranges contained more early successional cover (Clark 2000). Edges along access routes are a preferred cover of grouse in the Appalachian Mountains (Schumacher 2002, Tirpak 2005, Whitaker et al. 2005), providing abundant herbaceous groundcover and invertebrate foods, as well as grit for digestion (Hollifield and Dimmick 1995, Rusch et al. 2000). However, it has also been suggested that forest edges, such as those created by roadways, act as secondary habitat and are used more extensively when early successional stands are unavailable (Bump et al. 1947, Gullion 1984a, Whitaker 2003). This could account for the fact that adult male home-range sizes were more strongly influenced by clearcuts, while juvenile males responded more strongly to access routes, as territorial adult males can exclude juveniles from higher-quality early successional habitats (Small 1985).

Smaller spring–summer home ranges of females inhabiting oak-hickory forests contained a greater proportion of mesic (moist) bottomlands. This inverse relation suggests that these areas provided preferred brood-rearing habitat in oak-hickory forests. Other studies have found that grouse broods in southern portions of the species range select lower slope positions and riparian zones (Thompson et al. 1987, Fettinger 2002), particularly in oak-hickory forests (Stewart 1956, Whitaker et al. 2005). Researchers also have found that females with broods selected stands having well-developed herbaceous groundcover and high insect biomass (Haulton et al. 2003, Fettinger 2002), which together constitute the majority of the diet of ruffed grouse chicks (Rusch et al. 2000). Upland soils in oak-hickory forests typically are xeric and support little herbaceous understory vegetation, suggesting a reason for the association of smaller female home ranges with bottomlands. In contrast, uplands in mixed-mesophytic forests generally have mesic soils that support abundant understory herbaceous vegetation (Braun 1950), presumably relaxing this constraint.

A key goal of the ACGRP was to experimentally test the effect of hunting on ruffed grouse ecology; we found that grouse occupied smaller home ranges following closure of sites to hunt-

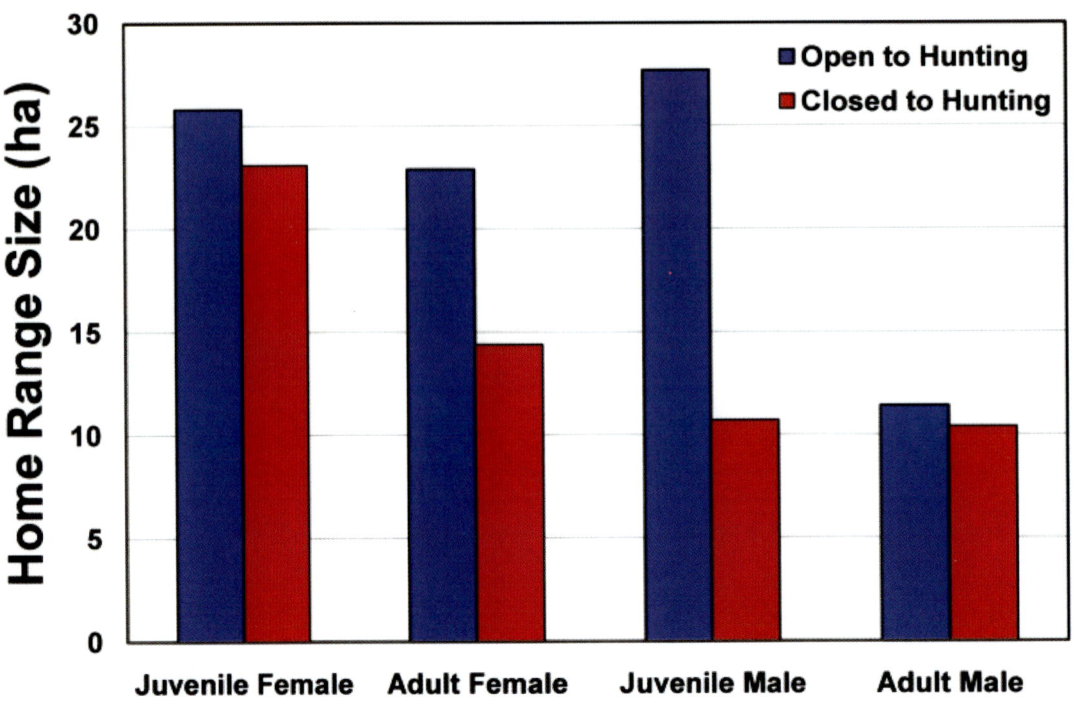

Figure 10.7

Mean home range size for ruffed grouse inhabiting three treatment sites during three years when hunting was open during fall and winter (1996–1999), and three years when hunting was closed (1999–2002).

ing (Figure 10.7). Presumably, grouse that are hunted may move more in response to hunter disturbance, resulting in a larger home range. This is in agreement with a study of hunting effects on ruffed grouse in Michigan (Clark 2000); increased movement in response to hunting pressure has also been reported for wild turkey and black ducks (Hoffman 1991, Clugston et al. 1994). We hypothesize that hunters, who typically focus their efforts on optimal habitat (Kilgo et al. 1998, Lyon and Burcham 1998, Brøseth and Pedersen 2000), may compel game to abandon these higher-quality areas in favor of lower-quality cover types. Subsequent evaluation of our data revealed that grouse made increased use of "preferred" covers following closure of hunting, offering support to this hypothesis (Whitaker et al. 2005). This response could lead to indirect negative effects of hunting on grouse, including reduced condition and increased predation rates, both of which could exacerbate consequences of hunting for populations. However, no significant change in predation rate or reproductive success was detected following closure of hunting (Devers 2005).

Our belief that a reduction in home-range size indicates an improvement in conditions rests on the assumption that grouse occupying smaller home ranges will have higher survival or reproductive success, resulting in higher fitness. We could not investigate the effect of home-range size on survival probability, though other studies have reported higher survival for ruffed grouse occupying smaller home ranges (Thompson and Fritzell 1989, Clark 2000). However, we did find that female grouse occupying smaller home ranges during the nonbreeding season were more likely to attempt to nest during the following spring (Figure 10.8). No relation was detected between fall–winter home-range size and nest success or brood success. Thus, though difficult to measure directly, the general premise that home-range size is inversely related to fitness appears to hold for ruffed grouse.

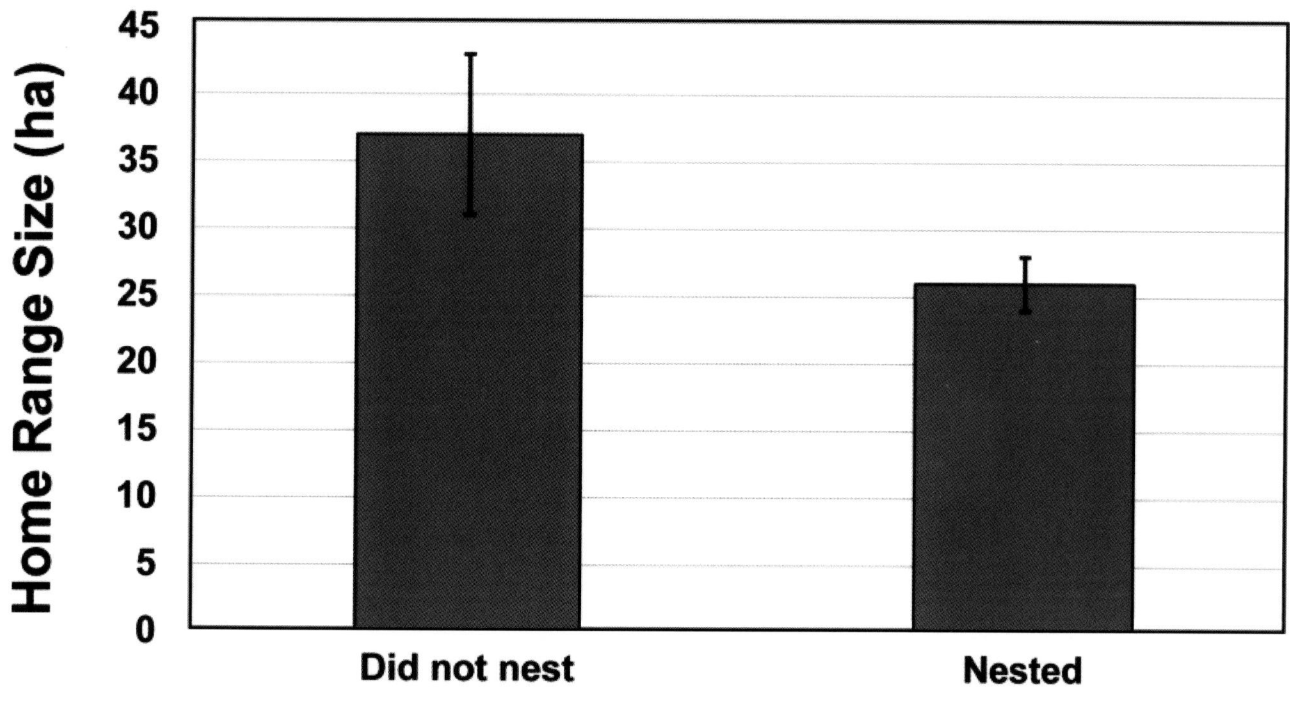

Figure 10.8

Comparison of mean fall-winter home-range size between female ruffed grouse that did not attempt to nest the following spring (n = 13) and those that did attempt to nest (n = 146) on ACGRP study sites, 1996–2001. Bars indicate one standard error.

Site Fidelity of Ruffed Grouse

When information is available on space use by an individual over an extended period, the practice of quantifying site fidelity is relatively straightforward. Perhaps the most obvious approach is to compare the location of home ranges used by an individual during different time periods. Because many of the ruffed grouse monitored by ACGRP co-operators survived for more than one season, we were able to use our radio-tracking database to quantify patterns of site fidelity of Appalachian ruffed grouse. Here, we considered fidelity only to sites occupied after natal dispersal, as we presumed that an individual must first have selected and settled in a site before site fidelity can be expressed or measured. Therefore, we used the location datasets from which dispersal moves had been removed.

A straightforward and useful measure of site fidelity is the straight-line distance between the centers of the home ranges used by an individual during the two time periods of interest (White and Garrott 1990). Home-range centers were estimated as the mean coordinates of all locations collected during the specified time period. Because we were interested in describing patterns of site fidelity over the lifetime of a grouse, we considered only those individuals for which we collected sufficient radio tracking data (i.e., at least 30 locations) to estimate home ranges for two or more years. We used these data to estimate site fidelity at three temporal scales: (1) we calculated the distance between pairs of home ranges collected in consecutive seasons (seasonal site fidelity); (2) the same season in consecutive years (annual site fidelity); and (3) for grouse captured during their first fall, between their first fall–winter home range and each subsequent home range (lifetime site fidelity). When a grouse had been captured as a juvenile we knew its precise age throughout its life. However, when an individual was captured as an adult we knew only that it was in at least its second fall–winter or second breeding season at the time of capture. Consequently, when we pooled different age classes of adult grouse we assumed that those captured as adults were in their second fall–winter or spring–summer at the time of capture. To recognize that some individuals may have been older than the specified age for a class, adult age classes are reported as, for example, fall–winter 2+. We numbered breeding seasons starting with the first year of maturity (i.e., first breeding opportunity), and in doing so disregarded the breeding season during which individuals hatched and were not yet independent of their parent (i.e., natal breeding seasons).

We were able to estimate seasonal dispersal for grouse up to fall-winter 5+. This reconstruction revealed contrasting patterns for males and females (Figure 10.9). Females consistently showed interseasonal home-range shifts of approximately 300 meters throughout their lives. In contrast, males showed a large shift in home-range location between their first fall–winter and first breeding season (mean = 533 m), and then declining seasonal shifts until their third fall–winter. From that point forward they showed relatively high site fidelity, with mean seasonal home-range shifts of approximately 85 meters, indicating that the location of male home ranges is generally stable from three years of age.

We were able to estimate lifetime site fidelity up to three years of age for individuals captured as juveniles (i.e., the distance to an individual's first fall–winter home range from each subsequent home range). Patterns over the lifetime of individuals indicated that grouse did not continue dispersing directionally away from their first fall–winter home range throughout their lives (Figure 10.10). This implies that shifts in home-range location over time were not directional.

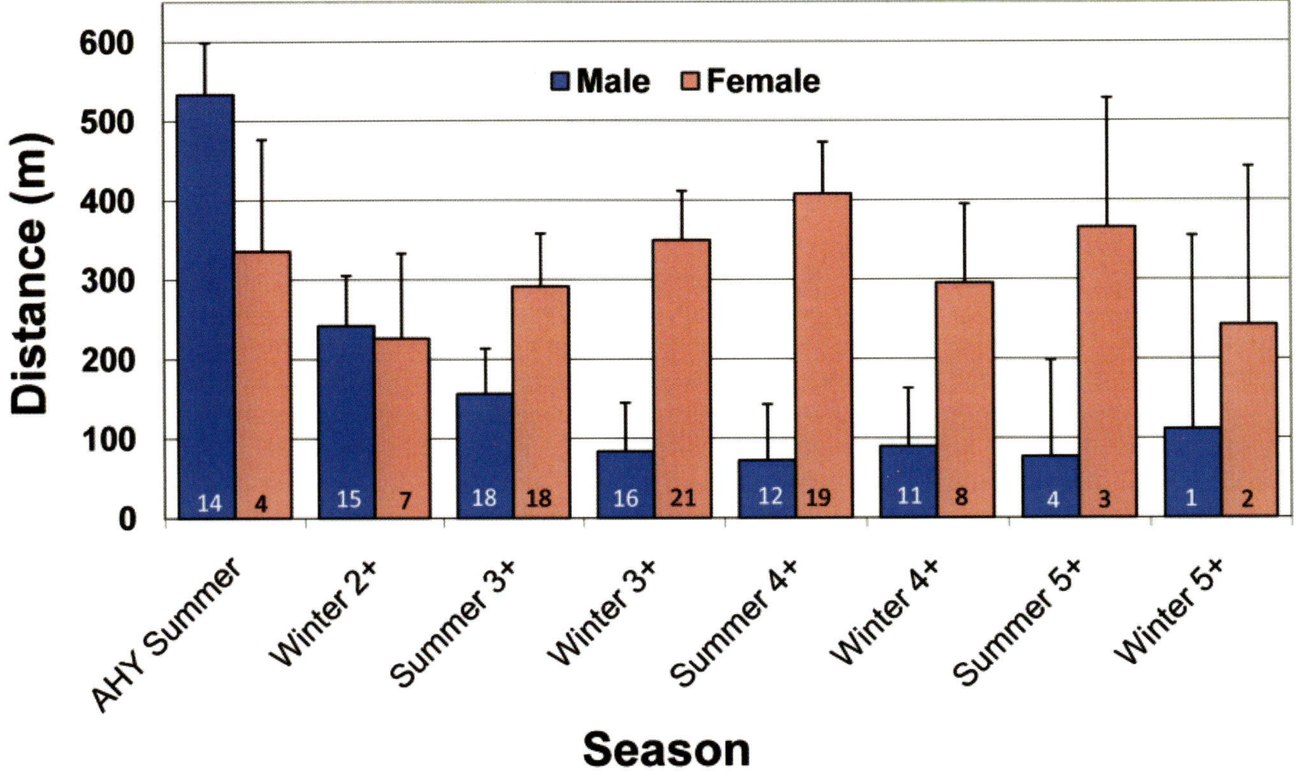

Figure 10.9

Seasonal site fidelity of Appalachian ruffed grouse, 1996–2001. Columns indicate the mean distance between the centers of the home range occupied during the current season and that used during the preceding season. Error bars indicate one standard error, while numbers on each column indicate sample size.

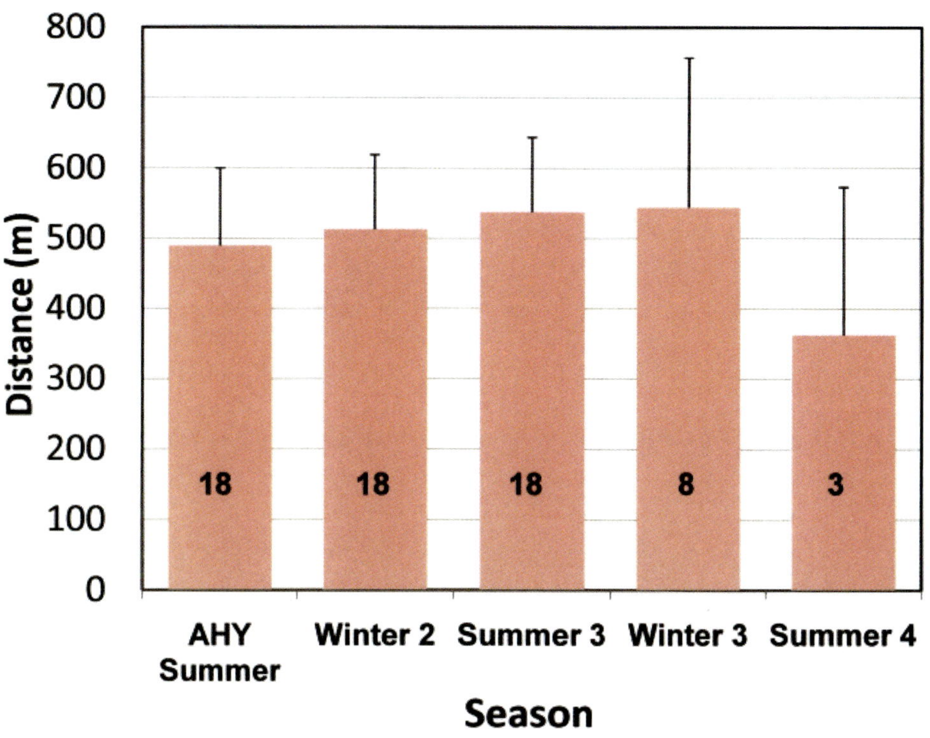

Figure 10.10

Lifetime site fidelity of Appalachian ruffed grouse, 1996–2001. Winter 2 represents the bird's second winter, Summer 3 the third summer, etc. Columns indicate the mean distance from the center of the home range occupied during the current season to that occupied during an individual's first fall-winter. Error bars indicate an estimate of one standard error, while numbers on each column indicate sample size.

Movements

The topic of movement is broad, touching on many aspects of a species' ecology including time budgets, energetics, predator avoidance, dispersal, and habitat use. In addition to potential implications for home-range size, daily movement rates are closely linked to time or activity budgets and may be affected by a number of factors. Along with having larger home ranges than grouse in northern forests, Appalachian ruffed grouse have been found to be active for approximately twice as much time each day during winter, and active throughout the day rather than being primarily crepuscular (active at dawn and dusk, Hewitt and Kirkpatrick 1997b). This may have implications for many aspects of the species' ecology, particularly energetics and encounter rates with predators, and, as with increased home-range size, may be indicative of low habitat quality. Elevated activity times and movement rates are most likely a consequence of low food availability or quality, adding to the body of evidence that grouse in the region suffer nutritional stress (Hewitt and Kirkpatrick 1997b).

We have suggested that movement increases the risk of encountering predators, and thus movement rates may be positively correlated with predation rates. In agreement with this, Thompson and Fritzell (1989) and Yoder et al. (2004) reported that predation rates were higher for grouse that made more extensive daily movements. However, it is important to recognize that some forms of movement are more dangerous than others. This may be particularly true of movements related to mating. Though grouse populations are at their annual low during the spring mating period, Bumann (2002) found that hourly sighting rates of ruffed grouse by observers on ACGRP study sites were highest at this time. Predation rates on grouse also were highest during this period, and if one assumes that detection rates by humans correlate with detection rates by predators, this suggests that grouse forgo stealth and undertake risky movements during the mating period. Yoder et al. (2004) also provided evidence that predation risk is in part a function of movement type, in that movement through unfamiliar areas carries a much greater risk of predation than movement in familiar areas. Site familiarity was of greater importance to predation risk than movement rate, providing an ecological explanation for why grouse should exhibit high site fidelity.

It is logical to predict that an increase in daily activity time will be reflected in increased movements by Appalachian ruffed grouse, though a lack of data for northern grouse precludes comparison at this time. Fearer (1999) studied daily movement rates by taking hourly locations on grouse on the VA3 study site. He found that, when averaged throughout the year, movement rates did not differ between males and females (mean of 109 meters), but that juveniles made more extensive hourly movements than adults (124 m/hour vs. 102 m/hour, respectively). Yoder et al. (2004) also reported that net daily movements of non-dispersing juvenile grouse are greater than those of adults (100m/day vs. 63 m/day, respectively). Other authors also have studied movement rates of Appalachian ruffed grouse. Epperson (1988) reported an hourly movement rate of 76 meters per hour and total daily movements of 946 meters for grouse in Tennessee during late summer. However, net daily movements of these individuals were only 161 meters, a reflection of the relatively high fidelity of grouse to their home ranges. Also related to site fidelity, Yoder et al. (2004) reported that in Ohio net daily movement rates were greater in unfamiliar areas for both juvenile (315 vs. 100 m/day; unfamiliar vs. familiar areas) and adult grouse (179 vs. 63 m/day). However it should be noted that this difference might, at least in part, be a consequence of the higher probability that grouse in unfamiliar areas are actively dispersing.

ACGRP researchers also found evidence that grouse movements were related to patterns of habitat use. Fearer (1999) found that hourly movement rates of grouse were positively related to the mean distance between habitat patches of the same type, as well as to percent cover of evergreen understory (mountain laurel or rhododendron). Conversely, movement rate was inversely related to the amount of high-contrast edge, and lower for grouse with home ranges having more than 25% cover of clearcuts (88 m/day) than those having less than 25% cover of clearcuts (119 m/day). These observations suggest that conditions are more favorable for grouse when landscapes have a high interspersion of small habitat patches, including a high percentage of clearcuts having hard, abrupt edges.

Haulton (1999) studied ruffed grouse broods on the VA2, VA3, and WV2 study sites in 1997 and 1998, and provided data on the distance of broods from nest sites over time. Broods sought out habitats having abundant groundcover vegetation and insect foods, and within one day of leaving the nest were 147 and 286 meters from the nest site (two broods), after two days were 202–679 meters away (four broods), and as much as 2,095 meters away after one week. One might predict that extensive movements by hens with broods may come at a cost to broods, and indeed chick survival to one week post-hatch was negatively related to distance moved from the nest (Tirpak 2000), suggesting that having a good interspersion of brood habitat in the landscape could lead to increased recruitment.

In summary, research conducted through the ACGRP sheds light on many aspects of the spatial ecology of ruffed grouse in the Appalachians. These findings are important to management, as factors affecting space use and movement also affect survival and reproduction of ruffed grouse, ultimately providing insight into factors limiting regional grouse populations. As found in other studies, grouse home ranges in the region were relatively large and were highly variable across sites and years. Much of this variability was related to local availability of key early successional escape covers, while in oak-hickory forests availability of moist bottomlands and hard mast were also important. Interspersion of habitats also affected space use, and findings suggested that landscapes having a high interspersion of small, symmetric habitat patches and an abundance of high-contrast edge were favorable for grouse. Finally, closure of hunting resulted in a reduction in home-range size of grouse, indicating that disturbance associated with hunting likely causes the birds to move more and have larger home ranges.

Appalachian ruffed grouse showed relatively high site fidelity, typically remaining within 500 meters of their first fall–winter home range for the remainder of their lives. This perhaps is not surprising for a species that is resident in an area year round. Males, which defend territories, showed increasing site fidelity up to three years of age, after which time distances between the centers of consecutive home ranges were on average less than 100 meters apart. In contrast, females, which do not defend territories, were much more mobile, typically shifting interseasonal home ranges by 300 meters throughout their lives.

Movement can also affect survival and reproduction, and ACGRP research identified several important aspects of movement patterns exhibited by Appalachian ruffed grouse. Landscape habitat structure and availability of key cover types affected movement rates of grouse, and, though costly in terms of chick survival, hens with broods often made extensive movements to gain access to brood habitat.

11 Dispersal

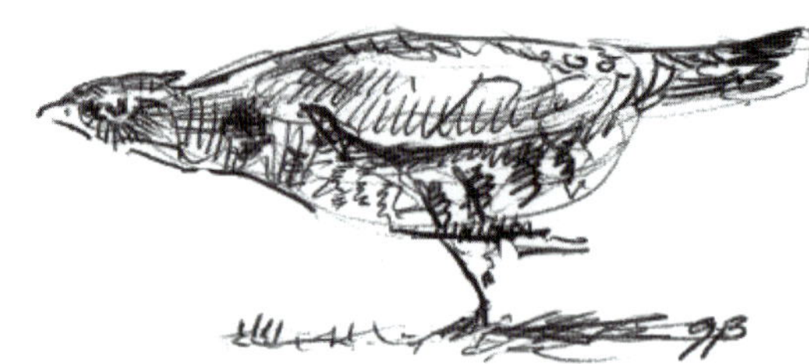

Brian W. Smith, Darroch M. Whitaker, and John W. Edwards

Dispersal in animals is the movement from where animals were born (or hatched) to either the first place they breed (called natal dispersal) or the movement between breeding sites (called breeding dispersal). Dispersal movements influence animal behaviors, population dynamics, conservation of the animals and their habitats, and even the evolution of natural populations. Clobert et al. (2001) identified three major components of dispersal: the decision phase (i.e., "stay or go"), transience phase, and a settlement (or colonization) phase. *Transience dispersal* is characterized by extensive movement from one area to another, whereas *colonization dispersal* occurs when an animal ceases extensive movement and attempts to establish a new home range (Johnson and Gaines 1987, Small et al. 1993). The risks encountered during each type of dispersal are likely different as well. Transient dispersers may be unfamiliar with the surrounding habitat, potentially exposing them to predators, or the area may be unsuitable habitat with limited food availability (Smith 1974, Wiggett and Boag 1989). During the colonization phase, an animal trying to settle in an area or secure a mate may compete with other animals already defending that territory (Garret and Franklin 1988, Nilsson 1989).

Understanding both breeding and natal dispersal is very important when trying to manage ruffed grouse populations. Many dispersal-related questions come to mind when resource agencies and researchers establish hunting seasons, purchase land for public use, and implement research or habitat management projects. For instance, what are some of the influences on dispersal timing, distances, and survival rates—mast production, occupancy of territories, available cover, etc.? What types of habitats are grouse using during transient dispersal? Are there mortality factors (e.g., predation) whose impacts can be reduced through habitat management? In this chapter, we discuss what we have learned about breeding and natal dispersal in Appalachian ruffed grouse.

Breeding Dispersal

Breeding dispersal, where animals relocate between breeding attempts, and site fidelity (see Chapter 10), the "loyalty" of an animal to a particular area, are two opposing behavioral concepts (i.e., should I stay or should I go?); however, both scenarios may offer advantages to ruffed grouse, depending on the situation. Fidelity to breeding sites offers familiarity with local habitat, so grouse would know where the best food, cover, and brood-rearing range were located. Site fidelity would also allow grouse to stay near previous mates. On the other hand, birds that choose not to move pass on the chance to find higher-quality breeding areas or mates (Greenwood and Harvey 1982). We examined two main questions related to breeding dispersal of ruffed grouse in the Appalachian Mountains: do they undergo breeding dispersal, and if so, what are some of the factors driving it?

Although patterns of natal and breeding dispersal have been described for ruffed grouse, causes of breeding dispersal have not been investigated. Ruffed grouse are polygamous (i.e., more than one female per male), they do not form "pair bonds," and males do not help to incubate eggs or care for young. Therefore, the competence of, or fidelity to, a mate should not affect whether a grouse stays at or near a breeding site (known as philopatry). Also, hens do not defend territories (Rusch et al. 2000), and will occasionally nest within sight of one another (D.M. Whitaker, personal observation), so it's doubtful that the presence of other grouse impacts nest-site selection very much. Female ruffed grouse do not typically reuse nest sites, but they generally occupy stable annual home ranges (Whitaker 2003) and sometimes nest as close as 50 meters from previous nest sites.

We assessed how age and nesting success might affect nest-area fidelity of ruffed grouse in the Appalachians. We analyzed nest-site data from September 1996 to September 2002 for 10 study sites (KY, MD, OH, PA, VA, and WV) in the project (Whitaker 2003). We considered a nest successful if at least one chick departed the nest with the hen. If we determined that the nest was depredated or abandoned, we considered the female a failed nester for that year. When nest fate was questionable, we designated the nest as successful if a one-week brood count yielded at least one chick, and considered the fate of the nest uncertain and eliminated the observation if no chicks were seen with the hen at one week post-hatch. We compared the straight-line distance between consecutive nest locations for an individual bird to examine nest-site fidelity. Because only about 8% of failed nesters attempt to renest in the region (Reynolds et al. 2000), these nesting attempts usually occurred in consecutive breeding seasons rather than within a given year. If a female renested after she lost her first nest, we calculated the distance between her first nest and second nest that same breeding season.

We evaluated nest-site fidelity in three different ways. First, we compared nest-site fidelity between yearlings (i.e., first time breeders) and adults, using only those hens that nested successfully during their first year and attempted to nest the second year. This comparison was designed to test for an effect of age on nest-site fidelity. Second, we compared internest distances between years for females who nested successfully in year one to within-year internest distances for females who lost a nest and then renested during that same breeding season. Third, we compared distance to the following year's nest between successful and unsuccessful nesters. The second and third tests were intended to look for effects of nesting success on within-year and between-year nest-site fidelity, respectively. We also tested whether the proportion of dispersers (for this study only, defined as "moves exceeding 0.5 km through an area that was not revisited later, to a new repeatedly traversed area;" Whitaker 2003) differed between successful and unsuccessful nesters.

So, what were we able to learn about breeding dispersal of ruffed grouse in the Appalachians? Considering those grouse that nested successfully, adult females nested twice as far away during the subsequent breeding season compared to yearling females (Table 11.1). Internest distances of females that renested immediately following nest loss (i.e., within the same year) were similar to internest distances for females that nested successfully in two consecutive years (Table 11.1). However, compared to those who had nested successfully, female grouse whose nests failed nested two and a half times farther away in the following year (Table 11.1). We (D.M. Whitaker, unpublished data) also analyzed home ranges of adult female ruffed grouse and found that 16 of 203 (8%) made fall breeding dispersal moves, and had settled by December 1 of the given year. Three of 151 females (2%) dispersed between late March and early April, just prior to nest initiation in the region. Of those individuals for whom nesting success during the preceding summer was known, failed nesters were more likely to disperse (4 of 15, 27%) than successful nesters (1 of 62, 2%).

Table 11.1. Breeding dispersal distances of female ruffed grouse in the Appalachian Mountains, 1996–2002. Three comparisons are presented: (1) for grouse that nested successfully, we compared distances to the subsequent year's nest between yearlings (i.e., first time breeders) and adults; (2) we compared distances to the subsequent year's nest for successful females to distance between nests for females who lost a nest and then re-nested during that same breeding season; and, (3) we compared distances to the subsequent year's nest between successful and unsuccessful females (D.M. Whitaker, unpublished data).

	Distance to Subsequent Nest (m)				
Group	*n*	Mean	Median	Range	Statistically Different?
Yearlings	14	167	124	38–441	yes
Adults	11	346	272	60–961	
Successful	25	246	175	38–961	no
Re-nested	15	333	300	96–666	
Successful	25	246	175	38–961	yes
Failed [a]	13	606	416	42–3046	

[a] Did not re-nest until the following year.

Both age and nesting success influenced nesting-area fidelity of female ruffed grouse. Adult hens that were successful had greater internest distances compared to successful yearlings; however, only one of 62 hens that had nested successfully subsequently moved to a new home range. Also, the distance between consecutive nests of adult females was relatively small compared to the typical home-range size of adult female ruffed grouse in the region (20–50 ha; Whitaker 2003). It appears that older females often nest farther from previous nest sites than do young females, but the adults nearly always remain within their existing home range. Small et al. (1993) found that a similar proportion of juvenile and adult female ruffed grouse were transient prior to nesting, but adults moved for a longer period of time than juveniles. Experience within a home range and/or age apparently play a large role in nest-site selection and fidelity; older birds might be more selective when choosing nest sites based on their previous experiences within their home range and familiarity with the habitat. Unfortunately, we do not know the complete breeding history of most adult females considered here; most of them were captured originally as adults, and hence it is impossible to determine how many nesting attempts these birds had experienced prior to entering the ACGRP. It is likely that many of these adults had experienced nest losses in previous breeding seasons but were successful during the first year we followed them. If this is true, then it is possible that previous nest losses continue to affect nest-site selection for more than one year.

Hens nested, on average, farther away the year after a failed nest (606 meters) than they did following successful nesting attempts (246 meters). Also, a greater proportion of birds that lost a nest (and did not renest until the following year) relocated to new areas during the nonbreeding period; thus, an increase in internest distance was associated with an increased chance that grouse would undergo breeding dispersal. However, the distances moved varied widely among females whose nests failed; some nested nearby the following year, whereas others moved up to three kilometers (Table 11.1). In contrast, females that renested immediately following a failed nesting attempt had internest distances similar to between-year internest distances for successful nesters (Table 11.1). Therefore, it appears that ruffed grouse breeding dispersal resulting from nest loss is delayed until the post-breeding season. Most breeding dispersal by adult female ruffed grouse occurs primarily during the nonbreeding period, particularly during fall. What determines how far a female is going to move (e.g., cause of nesting failure, individual variation, a subtle change in habitat) the year after a failed nest remains unclear.

Natal Dispersal

Most studies of natal dispersal (and survival during this period) in ruffed grouse have occurred in the northern portions of its range (Godfrey and Marshall 1969, Small and Rusch 1989, Clark 1996), where grouse occur in high densities and habitat differs markedly from the range in the central and southern Appalachians. In their northern range, researchers have found juvenile ruffed grouse to travel net average distances of 627 meters to 4,828 meters during fall dispersal movements, and that the habitat traversed is highly variable (e.g., regenerating forests, mature forests, forest edge, etc.). As noted earlier, many researchers believe that death is a risk inherent to dispersers. In ruffed grouse, Small et al. (1993) found no difference in survival between juvenile ruffed grouse undergoing transience and colonization dispersal (survival rates of 0.30 vs. 0.31, respectively) over the entire (autumn–spring) natal dispersal period in Wisconsin. They concluded that neither adult nor juvenile ruffed grouse were more vulnerable during dispersal overall, but emphasized the importance of distinguishing between transient and colonization dispersal.

Little information about ruffed grouse dispersal exists for their range in the central and southern Appalachians. In West Virginia and Ohio, respectively, Plaugher (1998) and Yoder (2004) found that juvenile grouse typically began dispersal in late September or early October, and averaged 2.4–3.7 km and 2.3–4.0 km (1.5-2.3 mi and 1.4-2.5 mi) traveled, respectively. One juvenile female followed by Plaugher dispersed 9.5 kilometers (5.9 mi) from its original trap site. Compared to northern studies, ruffed grouse in West Virginia began dispersal later and dispersed over a longer period of time, but were similar with regard to distances moved and randomness of directionality (Plaugher 1998). Plaugher's research laid the foundation for our further investigation of natal dispersal of juvenile ruffed grouse. We designed a project to investigate several factors that may influence dispersal behaviors, distances, and survival. Our objectives for this portion of the ACGRP were to address the following questions regarding dispersal of juvenile ruffed grouse:

1. What dispersal patterns do juvenile grouse throughout the Appalachians exhibit? Are there differences between sexes in various measures of natal dispersal (i.e., distances, movement and survival rates, and risks)?
2. Is mortality of juvenile grouse during dispersal associated with rates of movement or familiarity with a site?
3. What is the relation of juvenile dispersal and survival to forest type and acorns?

To address these questions, we analyzed telemetry data from over 1,200 individual juvenile grouse captured across 10 study areas during the ACGRP. Because peak juvenile dispersal occurs in the fall, birds that were captured in their first spring or early winter were excluded from this analysis—by the time these grouse were captured, they would have already dispersed if they had in fact chosen to disperse (not all juvenile grouse disperse). If birds were not located on a regular basis, dispersal movements were easy to miss. Therefore, if there was over three weeks between locations during September and October, we did not include those birds in all analyses because a short movement or temporary foray to an area with a subsequent return to their range could have gone undetected, or data would be lacking about the path traveled during dispersal.

We were interested in examining dispersal movements before first breeding attempts for juvenile grouse. Therefore, we used locations collected between grouse initial capture dates (allowing for the seven-day acclimation period) in late summer/early fall, until mortality or April 30 of the juvenile's first spring, whichever came first. We examined movement patterns and determined distances between home-range centers (if dispersal occurred). We determined the center of home ranges, and measured distances between home-range centers (or final location if the bird died during dispersal) after a dispersal event. Additionally, we identified short-duration "forays" taken by grouse outside of their 95% home range contours.

Transient dispersal events were identified as one-way movements at least 300 meters from an established home range (or capture area if there were too few points to delineate a home range before dispersal) in a nearly unidirectional manner for at least three consecutive locations. A grouse was considered to have dispersed if it moved over 300 meters from an established home range but never returned to that home range (Smith 2006). Ruffed grouse dispersal is often rapid, with daily movement distances of 100 meters to two kilometers reported in Ohio (Yoder 2004).

We classified juvenile grouse, based on movements, as: fall transients, winter transients, and non-dispersers. Fall transients (Figure 11.1) dispersed from their home range before November 15 of their first year, which encompasses peak fall dispersal for juvenile ruffed grouse in the region (Plaugher 1998, Rusch et al. 2000, Yoder et al. 2004). Winter transients (Figure 11.2) were juvenile grouse that dispersed after November 15 of their first year, leaving their established home range and not returning during the observation period. Several grouse exhibited both fall and winter transience and were therefore included in both of the categories above (Figure 11.3). Finally, non-dispersers (Figure 11.4) were juvenile grouse that did not leave their established home range during either fall or winter seasons. Although non-dispersing grouse did not permanently disperse from their home ranges, they often took short-duration (i.e., less than one-week) forays outside their home range. Forays appear to be a systematic approach for some species to search for suitable habitat in unfamiliar areas (Conradt et al. 2003).

Throughout the study period, ACGRP workers tracking grouse often lost contact with them for brief periods of time when dispersal events began, or some grouse moved so rapidly or extensively that we were unable to locate them for an extended period of time, if at all. For nearly all juvenile grouse, we had several locations for each bird near their original capture site, or within an easily defined home range, but they would sud-

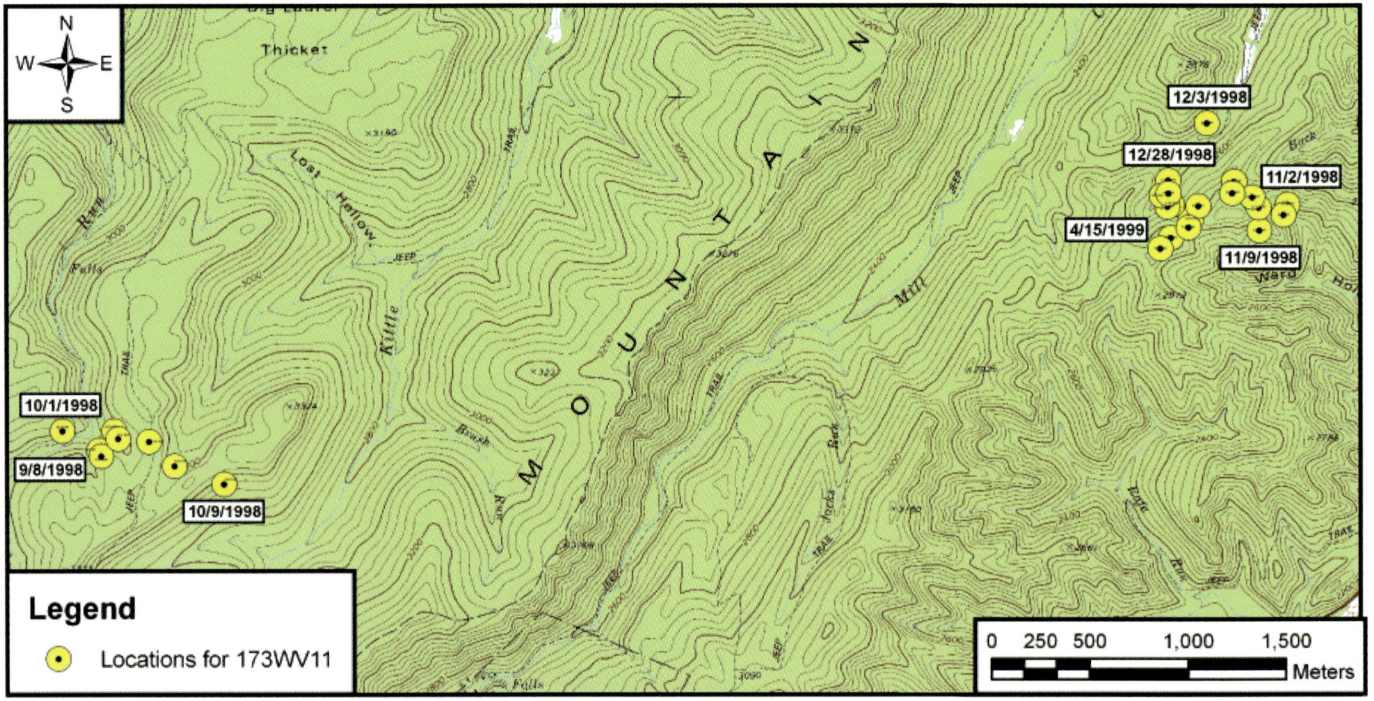

Figure 11.1

Estimated locations via radio telemetry for ruffed grouse 173WV1, a juvenile female originally captured in September 1998 on WV1 in Randolph County, West Virginia. Smith (2006) classified this grouse as a fall transient because she initiated dispersal prior to November 15. She dispersed eastward approximately 5,531 m and attempted to nest in her new home range in spring of 1999.

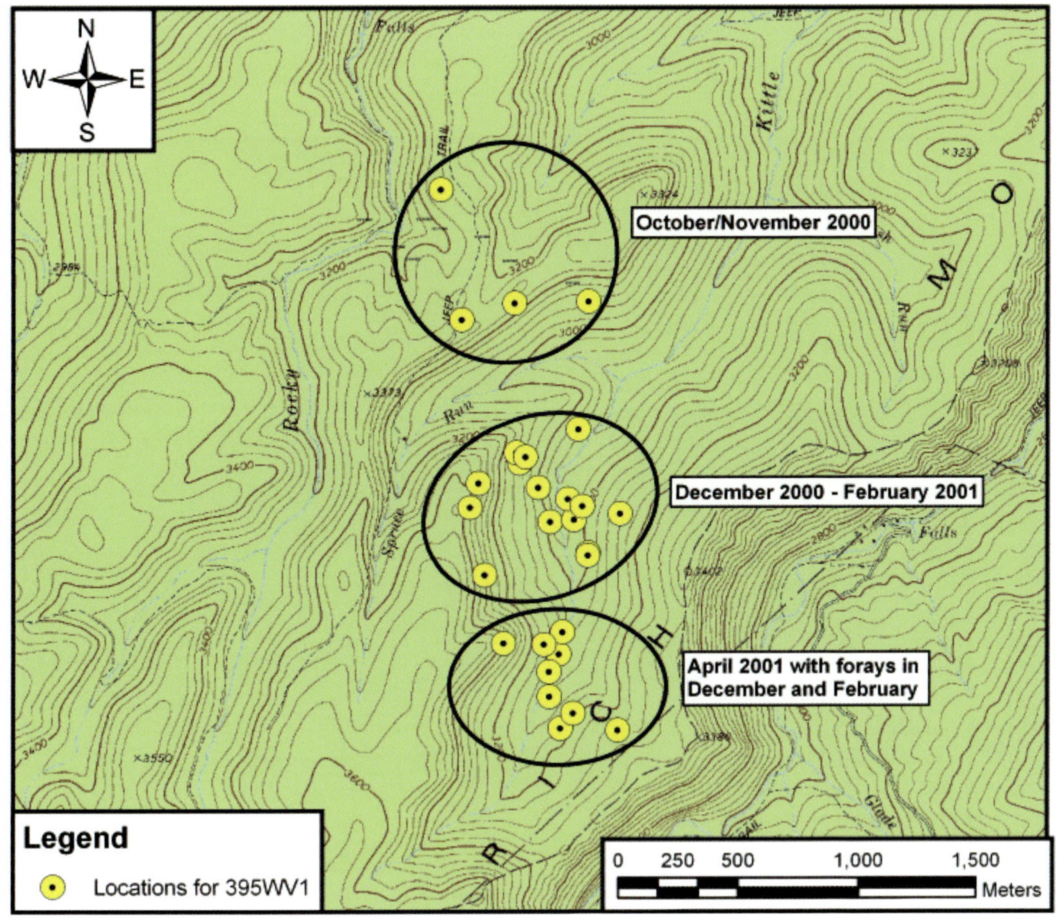

Figure 11.2

Estimated locations via radio telemetry for ruffed grouse 395WV1, a juvenile female originally captured in early October 2000 on WV1 in Randolph County, West Virginia. Smith (2006) classified this grouse as a winter transient because she initiated dispersal after November 15. She dispersed south approximately 1,123 m in early December, settling there until February/March 2001. During that time, she occasionally took short-duration forays to the area circled farthest south. In April 2001, she shifted her home range to the southernmost area circled and attempted to nest in the general area.

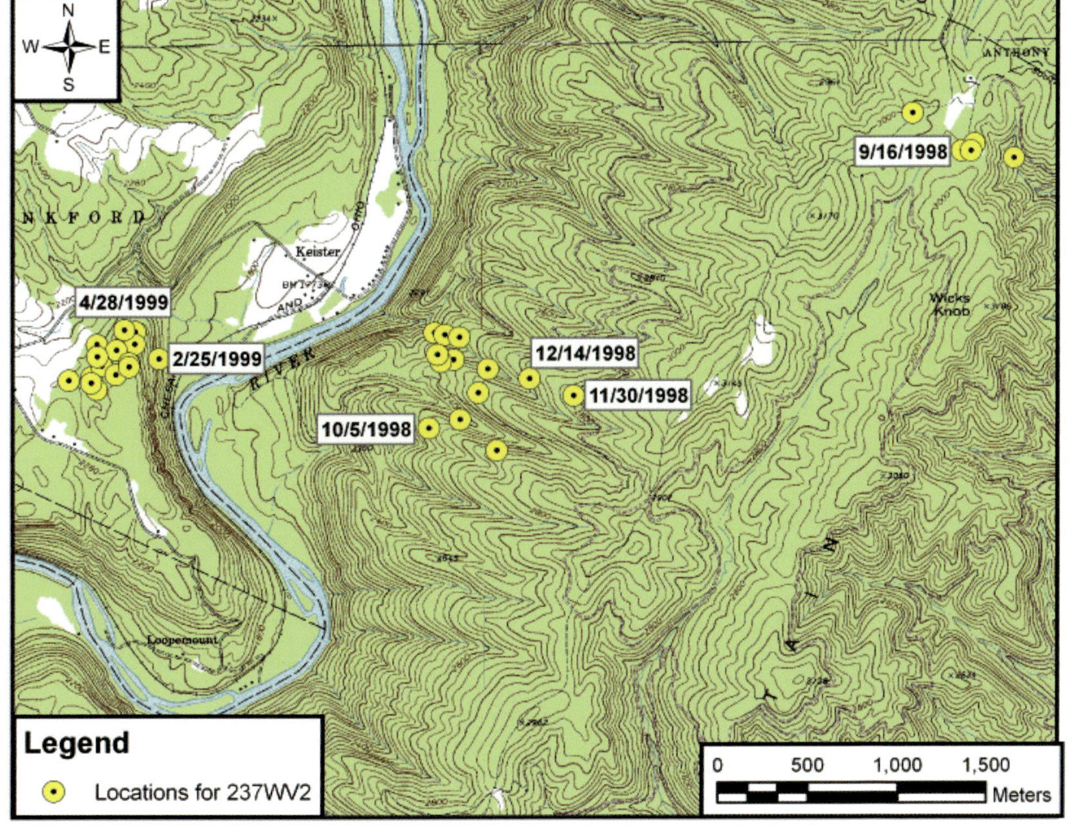

Figure 11.3

Estimated locations via radio telemetry for ruffed grouse 237WV2, a juvenile female originally captured in September 1998 on WV2 in Greenbrier County, West Virginia. Smith (2006) classified this grouse as both a fall transient and a winter transient because she exhibited two distinct dispersal movements: her first initiated prior to November 15 1998 (about 3,087 m) and her second began in late December 1998 (about 1,980 m). Female 237WV2 attempted to nest in her westernmost home range in spring of 1999.

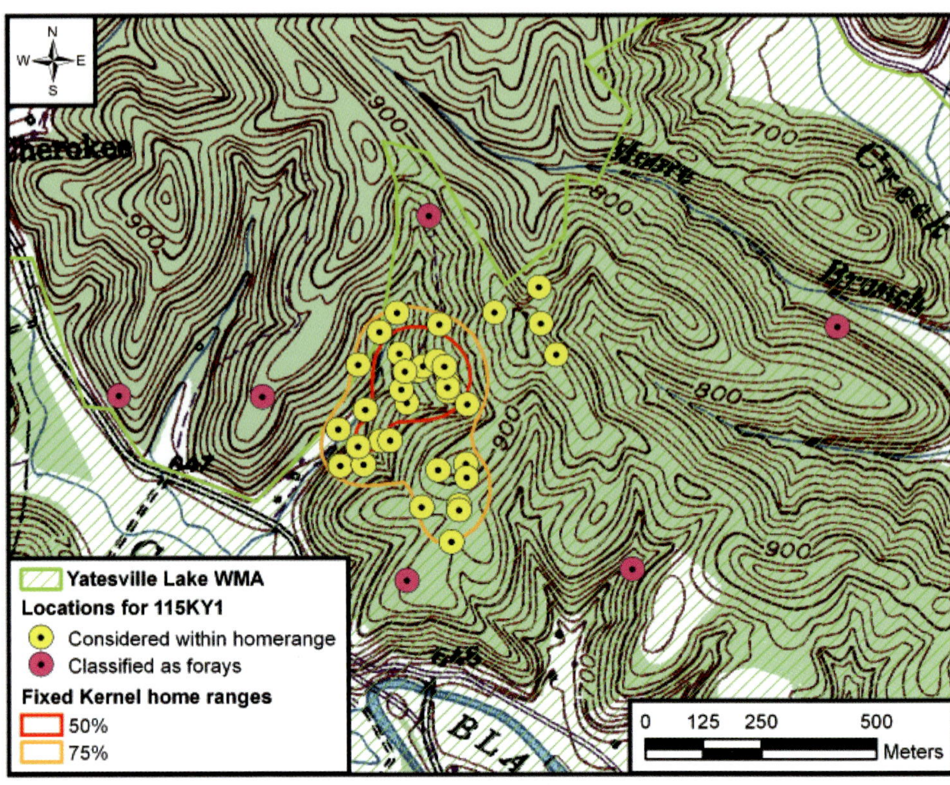

Figure 11.4

Estimated locations via radio telemetry for ruffed grouse 115KY1, a juvenile male originally captured in September 2000 on KY1 in Lawrence County, Kentucky. Smith (2006) classified this grouse as non-disperser because he exhibited no distinct dispersal movements in any season during the monitoring period. Locations highlighted by pink circles indicate short-duration forays because they either fell outside boundaries for the 95% fixed kernel home range, or they fell outside boundaries for the 75% fixed kernel home range and were (1) over 250 m from the center of the home range, or (2) over 150 m from the estimated 75% home-range boundary and isolated from all other points (i.e., surrounding locale only visited once). Grouse 115KY1 remained in this home range until his transmitter failed in October 2001.

denly "disappear" from their home range. Searches of the immediate vicinity, the entire study area, and/or from airplanes or vantage points outside the study area to locate these birds were conducted, often taking weeks and sometimes even months to relocate these individuals.

For the ACGRP, we were interested in influences on natal dispersal in grouse. We wanted to know if forest type (oak-hickory or mixed-mesophytic) and hard-mast production would influence dispersal distances and survival of grouse, similar to Whitaker (2003) who found that forest type affected home-range size after hard-mast crops failed. We also were interested in effects of grouse sex and disperser type (fall transient, winter transient, or non-disperser) on dispersal distances. We examined movement rates for all birds, regardless of fate, during different seasons. We calculated average daily rates of movement for grouse during the fall and winter periods. Finally, we examined if risk of mortality increases with movement distance or rate of movement. During dispersal, there are several time scales during which movement might influence the risk of mortality the most. Therefore, we analyzed overall mortality rate, the mortality rate two weeks prior to the last known location of the bird, and the rate between the final location and second-to-last location.

Table 11.2. Timing of dispersal events for juvenile ruffed grouse in the Appalachian Mountains that exhibited at least one dispersal movement, 1996–2002. For these birds, we had sufficient telemetry data to determine when grouse initiated and terminated dispersal movements. Fall dispersers initiated movement prior to November 15 in their first fall, and winter dispersers began their dispersal after November 15. Percentages in parentheses are not cumulative within each row; they indicate the proportion of juveniles from each sex class that dispersed.

Sex	Fall	Winter	Both [a]
Females	78 (85%)	14 (15%)	6 (7%)
Males	62 (89%)	8 (11%)	5 (7%)
All Juveniles	140 (86%)	22 (14%)	11 (7%)

[a] Juvenile grouse that dispersed in both fall and winter seasons.

A total of 285 juvenile grouse was classified as dispersers or non-dispersers, but season of dispersal (i.e., fall or winter) was determined for only 249 of these birds (128 females and 121 males). Thirty-six juvenile grouse (20 females and 16 males) dispersed sometime during the observation period, but we lacked sufficient location data for them to use in most analyses. For juvenile grouse in the ACGRP, at least one dispersal event was observed for 70% of all grouse (198 of 285). Nearly 86% (140 of 162) of juveniles for which we had reliable data underwent dispersal in the fall, 14% (22 of 162) dispersed only during the winter, and 7% (11of 162) of grouse dispersed during both fall and winter periods (Table 11.2). In relation to forest type, we collected dispersal information for 133 juveniles (79 females and 54 males) on mixed-mesophytic sites, and 116 juveniles (49 females and 67 males) on oak-hickory sites. We observed dispersal for 62% of grouse (82 of 133) on mixed-mesophytic sites and 69% (80 of 116) of grouse on oak-hickory sites. On mixed-mesophytic sites, 65% of females (51 of 79) and 57% of males (31 of 54) dispersed at least once, as compared to 84% (41 of 49) and 58% (39 of 67), respectively, for grouse on oak-hickory sites. Overall, proportions of males and females that dispersed during fall and winter seasons were similar on mixed-mesophytic and oak-hickory sites (Table 11.3).

Fall transients moved farther on average (2,525 m) than winter transients (1,424 m; Table 11.4). Female grouse tended to disperse farther than males, which is typical (Small and Rusch 1989), but we found that male grouse actually traveled farther on average during winter transience than did females (Table 11.4). Both mast index and forest type (mixed-mesophytic or oak-hickory) influenced dispersal distances; grouse tended to disperse farther in years of good hard-mast production. Of the several factors we evaluated that might influence movement rates of juvenile grouse, we found that only season influenced movement rates. Overall, grouse traveled about 423 meters per day; however, during fall their average daily

Table 11.3. Timing of dispersal events on mixed-mesophytic (M-M) and oak-hickory (O-H) sites for juvenile ruffed grouse in the central and south Appalachian Mountains that exhibited at least one dispersal movement, 1996–2002. For these birds, we had sufficient telemetry data to determine when grouse initiated and terminated dispersal movements. Fall dispersers initiated movement prior to November 15 in their first fall, and winter dispersers began their dispersal after November 15. Percentages in parentheses are not cumulative within each row; they indicate the proportion of juveniles from each sex*forest type that dispersed.

	Fall		Winter		Both [a]	
Sex	M-M	O-H	M-M	O-H	M-M	O-H
Females	43 (84%)	35 (85%)	8 (16%)	6 (15%)	1 (2%)	5 (12%)
Males	26 (84%)	36 (92%)	5 (16%)	3 (8%)	4 (13%)	1 (3%)

[a] Juvenile grouse that dispersed in both fall and winter seasons.

Table 11.4. Mean distance (m) traveled by male and female juvenile ruffed grouse during transience dispersal in fall (initiated prior to November 15 of any given year) and winter (initiated after November 15) from the Appalachian Cooperative Grouse Research Project, 1996–2002 (Smith 2006).

	Females			Males			Overall[a]	
Dispersal Type	*n*	Mean	Range	*n*	Mean	Range	*n*	Mean
Fall	77	2,857	374–9,534	61	2,105	422–9,635	138	2,525
Winter	18	1,097	413–3,754	13	1,876	440–9,238	31	1,424
Overall[b]	95	2,524	—	74	2,064	—	169	2,323

[a] Mean distances for both time periods regardless of grouse sex.
[b] Mean distances for both sexes of grouse regardless of when they dispersed.
[c] Mean distances for all grouse that dispersed in either fall or winter.

rates of movement increased to 501 meters per day and decreased in winter to 336 meters per day. In non-dispersing grouse, the average distance of 89 forays was 684 meters; males moved farther (726 m, n = 52) on forays than females (626 m, n = 37). Overall, dispersal distances in juveniles averaged over four kilometers (Table 11.5).

Table 11.5. Dispersal distances (km) of juvenile ruffed grouse in the central and southern Appalachians from 1996–2002. Sufficient telemetry data during their transience and/or colonization periods was lacking to allow additional analyses; therefore, only distances for periods when reliable data were gathered are reported.

	n	Median	Mean	Range
Fall Dispersal[a]	16	3.9	4.4	1.1–9.5
Winter/Spring Dispersal[b]	8	2.3	2.5	1.2–5.1
Effective Distance[c]	35[d]	3.8	4.6	9.1–12.7

[a] Dispersal events that began prior to November 15 in any given year.
[b] Dispersal events that began after November 15 in any given year.
[c] Straight-line distance from initial location or center of initial home range to the center of bird's ultimate home range (i.e., either at mortality or April 30). For example, a bird may have moved 1 km south in the fall from its initial location, but then moved again in the spring 1.5 km to the northeast; effective distance moved would be measured from the initial location to the center of the last home range, which in this case would be nearly due east from where the bird originated.
[d] Ruffed grouse 162NC1 (a juvenile male) actually had an effective distance of 0 km even though he moved 2.5 km in October of 2000; he returned in March to his initial home range producing a net effective distance of 0 km. Grouse 162NC1 was not included in calculations under this category.

We have illustrated the location data for grouse 113KY1, a juvenile male from Kentucky that was originally captured in August of 2000, to illustrate the challenge in relocating grouse that had dispersed (Figure 11.5). This male was relatively easy to locate initially, even though he stayed just outside the study area; then in early September, he essentially vanished. On October 19, 2000, he was relocated over four kilometers away from his initial locations, in a seemingly random direction and across some very rugged terrain. This was the only time he was found there or anywhere in the immediate vicinity, notwithstanding extensive searching. Male 113KY1 was not relocated until January 24, 2001, when he was over three kilometers away from his October location. Additionally, he had crossed the Big Sandy River (a medium-sized river) into West Virginia. From his initial locations to where he settled in West Virginia was about seven kilometers, and he had not traveled a straight path, nor given any indications as to when he was starting dispersal and which way he was headed.

Mortality risks were lower for juveniles on mixed-mesophytic study areas than on oak-hickory sites. Familiarity with a general location apparently influenced mortality risk; risk of death for grouse in unfamiliar space was 2.3 and 2.4 times greater than for grouse in familiar space during the last two weeks or over the last two locations, respectively. Overall rate of movement also increased mortality risk, but the rate of movement during the last two weeks of life or between the last two locations had no apparent effects on risk. Similarly, the sex and acorns apparently had little influence on mortality risks for juvenile ruffed grouse in either of the sets of analyses, although we observed that male grouse generally experienced a slightly greater risk of mortality than females during dispersal.

Overall, 70% of juvenile grouse in the ACGRP exhibited at least one dispersal movement; this estimate probably was conservative because some of the "non-dispersing" juveniles may have been misclassified. Most grouse in the region initiate dispersal during the first two weeks of October (Plaugher 1998, Yoder 2004), but some begin dispersing during the second and third weeks of September. Therefore, it is likely that some, or many, of the "non-dispersers" were captured after they already had completed dispersal in the fall. With this in mind, it appears that dispersal may be much more common in the central and southern Appalachians than in northern portions of ruffed grouse range. Clark (1996) found that only 48% of juvenile ruffed grouse in northern Michigan dispersed in the fall, although this was likely a minimum estimate because she also included birds captured in October in her analyses.

Similar to Yoder (2004), who found that 75% of juvenile grouse dispersed in the fall in Ohio, we found that most (86%) juvenile grouse that dispersed did so in the fall. Why do ruffed grouse exhibit a tendency to disperse in the fall? Inbreed-

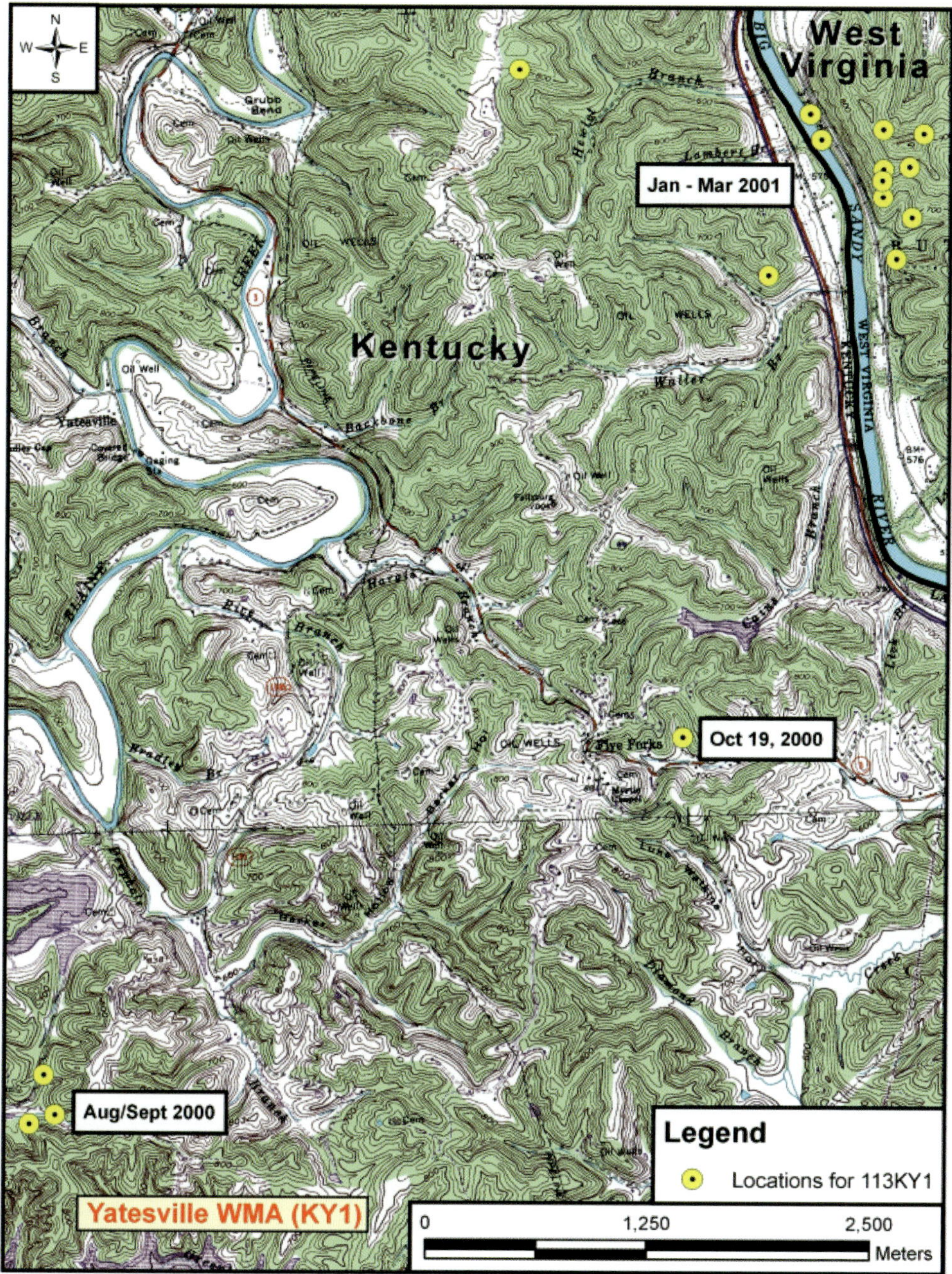

Figure 11.5

Estimated locations via radio telemetry for ruffed grouse 113KY1, a juvenile male originally captured in August of 2000 on KY1 in Lawrence County, Kentucky. Although this grouse initiated dispersal prior to November 15, Smith (2006) lacked sufficient telemetry data to include him in most of his analyses. However, Smith (2006) was able to determine that grouse 113KY1 had an effective dispersal distance of over seven kilometers, and he remained in the area where he settled until he was legally harvested in November 2001.

ing avoidance, local inter- and intra-specific competition, and competition among kin have been identified as important factors influencing natal dispersal in other species (Greenwood and Harvey 1982, Clarke et al. 1997, Lambin et al. 2001). In ruffed grouse, there are typically large-scale movements by juveniles shortly after broods break up in the fall (Godfrey and Marshal 1969, Small et al. 1991), and this may lead to immediate competition for resources with kin, unrelated grouse, or other species in the area.

Interestingly, winter dispersal patterns in the ACGRP were more similar to those of northern grouse populations than what Yoder (2004) observed in Ohio. He found that 43% of juveniles dispersed in spring (note: our winter period encompassed his spring period), but in Wisconsin, Small et al. (1993) observed 24% of juveniles dispersed in the winter and spring seasons combined (pooling sexes and seasons). In our study, 20% of juveniles dispersed in winter (i.e., birds that dispersed only in winter plus birds that dispersed in both fall and winter). Winter dispersal rates for male grouse on mixed-mesophytic sites were nearly three times greater than those of males on oak-hickory sites, and rates for females were one and a half times greater on oak-hickory sites than on mixed-mesophytic sites. Small et al. (1989) suggested that natal dispersal in male ruffed grouse may not be complete until spring, and that competition among males for potential breeding territories is higher than females for nesting areas. Devers (2005) found that productivity and recruitment were higher on mixed-mesophytic sites than on oak-hickory sites in the ACGRP, which suggests a greater potential for competition with other juvenile grouse on mixed-mesophytic sites while searching for a location to colonize. In female grouse, higher rates of winter dispersal on oak-hickory sites may be influenced most by the ephemeral nature of hard-mast crops and nutritional constraints associated with mast failures (e.g., Whitaker 2003, Norman et al. 2004, Devers 2005) or searches for preferred habitat (e.g., mesic bottomlands; Whitaker 2003) for upcoming nesting and brood-rearing seasons.

We also observed that juvenile female ruffed grouse in the ACGRP dispersed farther than males on average; this pattern is fairly typical for grouse (Keppie 1979, Hines 1985, Small and Rusch 1989) and for birds in general (Greenwood 1980). Average dispersal distances in fall for juvenile grouse in the ACGRP were similar to other studies (Small and Rusch 1989, Plaugher 1998), and longer movements occurred during the fall season than in winter. Small and Rusch (1989) suggested that extensive, wandering movements of males in late winter/early spring may be caused by competition among males for potential breeding territories, whereas competition among females for nesting areas may not be as intense, as they settle in an area most frequently in fall or occasionally winter.

In fall, forays likely allow juveniles to explore new areas for potential refugia in winter months (e.g., localized food resources, dense cover) or vacant territories, subsequently returning to a familiar area to feed more efficiently or avoid predation (Conradt et al. 2003). In winter or late spring, forays by juvenile grouse are likely related to searches for vacant drumming areas or potential mates, similar to "wanderings" of spring dispersers (Small and Rusch 1989). We occasionally observed juvenile grouse that had taken previous forays to eventually disperse to the area visited on the foray. Conradt et al. (2003) suggest that foray search dispersal is more efficient than random dispersal, allowing animals to initially concentrate in familiar areas and gradually expand outward and return to a familiar area; subsequent searches can then be based on information gathered in previous forays.

In terms of mortality risk during dispersal, forest type seemed to play an important role; juvenile ruffed grouse on oak-hickory sites were 1.3–1.7 times more likely to die during dispersal than those on mixed-mesophytic sites (Smith 2006). Conditions on oak-hickory sites may be more extreme for grouse, likely because these sites are more xeric and offer fewer northern hardwood trees (e.g., aspens, birches, and cherries) and sparser understory vegetation (Whitaker 2003). Therefore, mortality of dispersing grouse on oak-hickory sites is likely influenced by a complex interaction of hard-mast production and interspersion of high-quality habitats. Similar to Yoder et al. (2004) in southeastern Ohio, we also found that grouse in an unfamiliar area had an increased risk of mortality as compared to grouse in a familiar area. Moving through unfamiliar space could decrease foraging efficiency and success in avoiding predators, thereby possibly leading to apparent increase in mortality risk we observed for juvenile grouse in unfamiliar areas.

Through the ACGRP, we determined that many aspects of dispersal (both breeding and natal) for ruffed grouse in the central and southern Appalachian Mountains were similar to ruffed grouse throughout their range and many other gallinaceous birds. Most notably, breeding dispersal seems related to age and nesting success, and forest type (oak-hickory vs. mixed-mesophytic) influences various aspects of natal dispersal in ruffed grouse. We found that dispersal distances and mortality risk during dispersal varied by forest type for both male and female grouse. Additionally, dispersing through unfamiliar space increased mortality risk for ruffed grouse.

12

Harvest Management

Patrick K. Devers, Gary. W. Norman,
David A. Swanson, and Dean F. Stauffer

Regulated recreational hunting in the United States grew out of the crusades of sportsmen during the 1800s to stop market hunting and trade in wildlife parts that was causing the wholesale slaughter and decline of wildlife across the young nation. Many of these early sportsmen's groups were outraged by the unethical harvest and demise of white-tailed deer, bison, shorebirds, and the heath hen (a close relative of the ruffed grouse). Sportsmen and conservation groups like the New York Sportsmen's Club, the Boone and Crockett Club, and state chapters of the Audubon Society lobbied for and gained the passage of game laws and ensured their enforcement in their local communities (Trefethen 1975). A prominent concern of many of these groups was the harvest of upland gamebirds during the spring reproductive season. In fact, the unsustainable harvest of ruffed grouse in the spring led the New York Sportsmen's Club to introduce and obtain passage of a law in 1846 that banned the hunting, possession, and sale of ruffed grouse between the first of January and September (Trefethen 1975).

The formation, evolution, and success of early sportsmen and conservation groups laid the foundation for the field of wildlife conservation and the establishment of wildlife agencies throughout the nation. From the earliest beginning of the wildlife movement, the management of harvest for meat, feathers, or other products was of primary concern (Trefethen 1975). As public support for regulating the harvest and trade of wildlife grew, so did the foundations for regulated sport harvest. During the last years of the nineteenth century and the early years of the twentieth century, the current system of state wildlife management and enforcement of game laws took shape. The success of this system was due in large part to the innovative idea of T. Gilbert Pearson, who first proposed charging a hunting license fee and using the funds to administer and enforce game laws. With a secure source of funding, state wildlife agencies were able to effectively enforce state game laws, and enabled professional biologists to pursue the scientific management of our wildlife resource.

In the early part of the twentieth century, harvest management was akin to steering a freighter on the high seas. Harvest theory was limited or non-existent, and biologists had few tools to monitor and manage harvest. State wildlife agencies relied on bag limits and season structures to prevent over-harvest. Harvest management at the time seems to have been based on the assumption that hunting pressure and success would decrease as game populations decreased (Strickland et al. 1994). The underlying assumption of this argument is that each animal harvested by a hunter was in addition to the number of animals that were killed by natural causes, particularly predation, and could lead to a decrease in game abundance. This theory is known as the Additive Mortality Hypothesis (Chapter 5, Figure 12.1). However, in 1933, a researcher named Paul Errington proposed the Compensatory Mortality Hypothesis (Errington and Hamerstrom 1935). Based on his work with northern bobwhite quail, Errington posited that a piece of land could support (i.e., provide shelter and food) a limited number of animals during the harshest season of the year (usually winter in North America). This limit was referred to as the carrying capacity of the land. During spring and summer, populations would exceed the carrying capacity and produce excess animals. This excess was termed the doomed surplus, and Errington argued they would die from a combination of natural causes throughout the winter. By the following spring, the population would return the equilibrium with the carrying capacity of the land. In other words, the compensatory mortality hypothesis states the magnitude of each mortality factor (e.g., avian predation, mammalian predation, disease, starvation) acting upon a population may vary over time, but total mortality (usually measured over a one-year period) remains constant (Fig. 12.1). This hypothesis forms the foundation of regulated sport hunting throughout the United States (Strickland et al. 1994) and assumes human harvest of wildlife replaces one or more natural mortality factors (e.g., predation) without reducing the breeding population. This hypothesis generated great debate among biologists and managers, and spurred numerous studies to test its validity. These studies often produced ambiguous results, some supporting Errington's hypothesis and others refuting it. In 1976, researchers studying the influence of harvest on waterfowl proposed a third hypothesis known as the Intermediate Hypothesis (Anderson and Burnham 1976; Figure 12.1). This hypothesis states that harvest is compensatory up to a threshold, above which it becomes additive. This hypothesis strikes a balance between the Compensatory and Additive hypotheses and may explain why previous studies produced conflicting results.

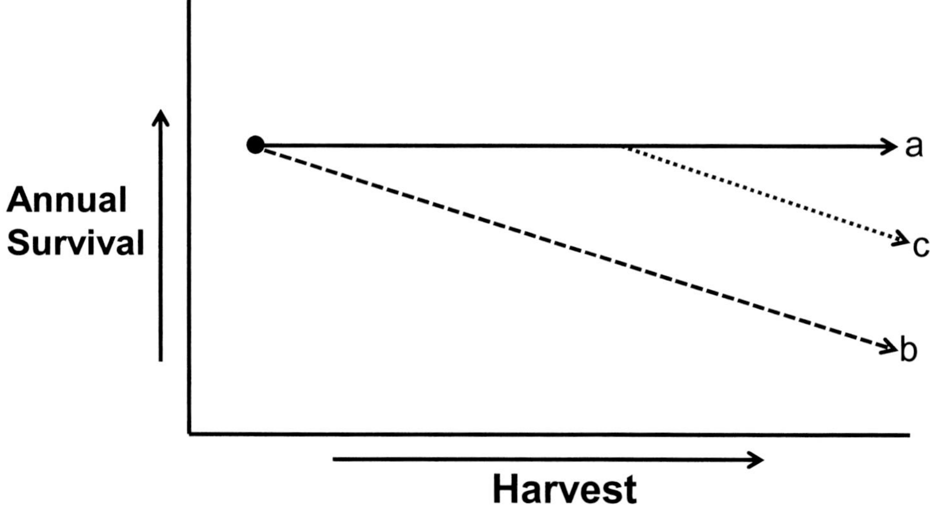

Figure 12.1
Theoretical predictions of the effect of harvest on annual survival under the (a) compensatory mortality hypothesis, (b) additive mortality hypothesis, and (c) intermediate hypothesis.

Ruffed Grouse Hunting

Throughout its range, the ruffed grouse is a very popular gamebird and biologists and managers have committed a great deal of effort to understand the effects of hunting on grouse populations. Several authors have concluded sport harvest does not negatively affect ruffed grouse populations (Bump et al. 1947, Dorney and Kabat 1960, Gullion and Marshall 1968, Fischer and Keith 1974). In one study, researchers estimated grouse abundance in small woodlots of western North Carolina with prescribed hunting pressure (no hunting, moderate hunting, and unrestricted hunting) before, during, and after hunting season. The researchers found grouse abundance did not differ among woodlots or after hunting, suggesting harvest mortality was compensatory (Monschein 1974). In Wisconsin, researchers concluded harvest mortality of <40% of the preseason population is compensated for with changes in natural mortality rates (Dorney and Kabat 1960). Similarly, other researchers have concluded harvest rates of <50% of preseason ruffed grouse populations are compensatory (Palmer 1956, Gromely 1996, Clark 2000). In a recent study in Ohio, harvest accounted for only 8.6% of known mortalities and was determined not to affect the grouse population (Swanson et al. 2003).

On the other hand, Bergerud (1985) posited all grouse hunting is additive to natural mortality and reduces spring breeding populations. In central Wisconsin, harvest rates of 28–33% were determined to be partially, if not completely additive and on a regional basis could cause a reduction in the abundance of grouse numbers (DeStefano and Rusch 1986, Small et al. 1991). Kubisiak (1984), also working in Wisconsin, drew a similar conclusion, stating a mean harvest rate of 44% (range 23–72%) was additive to natural mortality and caused a decline in grouse densities. He further argued that heavy harvest in the *early* part of the season could reduce spring grouse numbers (Kubisiak 1984). Interestingly, or perhaps just more perplexing, most biologists argue the proportion of juvenile grouse in the hunter's bag decreases as the season progresses, suggesting *late-season* harvest may reduce spring breeding populations by removing adults from the population (Dorney and Kabat 1960). Based on this conclusion, researchers have argued it is advantageous to harvest populations early in the fall to allow more time for compensatory processes to act on the population (Baines and Linden 1991).

Still yet, other research has supported the intermediate hypothesis. During their classic study of ruffed grouse ecology in New York, Gardiner Bump and his colleagues tested the Compensatory Mortality hypothesis. This group of pioneering grouse biologists experimentally harvested 19.5%, 20%, and 13.4% of the fall population on one study area and compared over-winter survival to a reference area over three consecutive years. They found over-winter survival was 45.2%, 55.8%, and 65.8% on the hunted area, which compared favorably to over-winter survival rates of 39.1%, 43.4%, and 60.5% on the reference area (Bump et al. 1947). The researchers concluded that decreases in natural mortality rates could compensate for 50% of harvest mortality, anything above 50% would be additive and would leave empty drumming logs and coverts the following year (Bump et al. 1947). Biologists studying willow ptarmigan (a majestic bird in its own right, but not the King of Gamebirds) recently completed a very similar study in Norway to test the Compensatory Mortality hypothesis. The researchers experimentally harvested 0%, 15%, and 30% of willow ptarmigan populations on 13 study areas during a four-year study and found that changes in natural mortality rates could compensate for 33% harvest rate (Pedersen et al. 2004). In central Wisconsin, ruffed grouse harvest mortality was higher on public than private land for juveniles (0.56 vs. 0.09, respectively) and adults (0.73 vs. 0.13, respectively), yet survival during the non-hunting season was similar (0.80 vs. 0.77) indicating harvest mortality was only partially compensatory on public lands (Small et al. 1991).

So, what is the effect of hunting ruffed grouse in the central and southern Appalachians? This was a primary question of the ACGRP. To answer this question, we used an experimental test to compare changes in ruffed grouse survival between hunted and non-hunted areas; we have provided the details of this experiment in Chapter 5. The ACGRP hunting experiment was conducted on seven study areas during the six-year study. During Phase I (fall 1996 to summer 1999) each of the seven study areas was open to normal hunting season following state regulations. Hunting seasons typically ran from early October to late February with daily bag limits ranging from one to four birds and possession limits of four to eight birds. During Phase II (fall 1999 to summer 2002) three treatments areas (KY1, VA3, and WV2) were closed to hunting while the four remaining control areas remained open to normal hunting seasons and regulations. This design provided ACGRP researchers a unique opportunity to test the Compensatory Mortality Hypothesis.

During the course of the ACGRP, hunters harvested 117 radio-collared birds (including legal harvest, wounding loss, and illegal harvest). Birds were harvested during each month of the season. The average harvest rate on control areas (i.e., those open to hunting each year) was only 8% (range 5–12%). In comparison, the average harvest rate on the three treatment areas between fall 1996 and winter 1999 was 20%. Harvest accounted for 12% of all known mortalities. Harvest rates experienced during the ACGRP were much lower than reported for other parts of ruffed grouse range, which may reach 29–50% (Palmer 1956, Dorney and Kabat 1960, Palmer and Bennet 1963, DeStefano and Rusch 1986, Small et al. 1991). Over the six-year period, ruffed grouse annual harvest did not differ between control and treatment areas, even after hunting was closed on the three treatment areas (Figure 5.3). Cessation of hunting on the three treatment areas did not result in increased survival among adult, juvenile, female, or male ruffed grouse. These findings indicate harvest rates of <20% are compensatory. However, we stress our results merit caution in concluding harvest mortality is generally not additive at all levels of hunting; *higher harvest rates may be additive*. We believe current harvest rates in the states studied can be maintained, but regional state agencies should not amend hunting seasons to facilitate higher harvest rates. It is critical to recognize our results are not conclusive due to limitation in sample size and effect size; we cannot assume harvest rates higher than those observed in this study are compensatory, nor can we extrapolate our results beyond the central and southern Appalachian region.

Although we believe regulated sport harvest did not have a direct impact on ruffed grouse survival, there is evidence that disturbance from hunting (and other activities) influenced habitat selection and home range size of ruffed grouse in the Appalachian region (Whitaker 2003). Ruffed grouse (regardless of sex and age classes) made greater use of clearcuts and mesic bottomlands and had smaller home ranges in the absence of hunting (Whitaker 2003). We believe this type of disturbance deserves consideration in the development of ruffed grouse hunting regulations and land management plans.

Recommendations

Our results indicate adult ruffed grouse in the central and southern Appalachian region experience high survival rates and that current harvest rates (<30%) are sustainable. Yet, other research has indicated that disturbance from hunting (and other sources) including vehicle traffic and flushing can cause changes in animal behavior, physiology, habitat selection, and potentially population dynamics (Knight and Cole 1995, Whitaker 2003). Time and again, research has demonstrated that hunting pressure, harvest rates, and hunter-related disturbance are related to distance from roads or initial starting points (e.g., gate or hunting cabin; Fischer and Keith 1974, Gullion 1983, Broseth and Pedersen 2000, Gratson and Whitman 2000, Hayes et al. 2002, McCorquodale et al. 2003). In Maine, Michigan, and Wisconsin, the majority of ruffed grouse hunting occurs within 400 m (quarter mile) of roads (Gullion 1983). In Alberta, harvest rate (48%) was higher for birds trapped <101 m from a road than birds trapped >101 m from the road (19%; Fischer and Keith 1974). In Norway, harvested willow ptarmigan lived closer to hunting cabins and had up to twice the amount of hunting pressure in their home ranges than individuals that survived the hunting season (Broseth and Pedersen 2000).

In light of these findings, we recommend state agencies manage ruffed grouse hunting in the central and southern Appalachian region at current harvest levels and for high quality experiences. To provide high quality hunting opportunities (i.e., low hunting pressure, low vehicle traffic, high flush rates), we recommend using road closures in conjunction with habitat management. In areas identified specifically for ruffed grouse management, we encourage closing roads from the start of the hunting season until the end of the early brood period (late June to mid July). Closing roads during this period will decrease disturbance during the two most critical periods of the year for ruffed grouse (i.e., winter and the breeding season). In areas managed for multiple use, and particularly areas that experience high levels of hunting for other species, we strongly encourage closing roads in the late hunting season (i.e., mid December) to the end of the early-brood period. This strategy should provide road access to hunters during archery, muzzleloader, and rifle seasons, but minimize disturbance to ruffed grouse during late winter and the breeding season. Further, we recommend ruffed grouse management units be divided into "refuge" and "recreational" areas.

Refuges are defined as areas that receive habitat management (i.e., clearcuts, selective cuts, prescribed burning) but are located >400 m (quarter mile) from a road or other access point (Fig. 12.2). Refuge areas will minimize recreational disturbance on ruffed grouse during critical times of the year (i.e., late winter and spring), allowing them to reduce their home range size and make greater use of preferred habitat features (e.g., regenerating clearcuts, access routes, and mesic bottoms: Whitaker 2003). It is possible that refuge areas will produce birds that will disperse across the landscape and may be available to hunters in recreational areas. Besides, any good grouse dog knows the best coverts are found far from the road in areas where few hunters have trampled, usually down the hollow, across the creek and just up the other side of the hill in the grape brambles. With a good dog in the lead and a willingness to enjoy a short hike during a cool, crisp winter day grouse hunters will find these refuge areas a paradise on earth.

Recreational areas are defined as areas that receive habitat management and are within 400 m (quarter mile) of a road or other access point (Fig. 12.2). These areas will provide great habitat for ruffed grouse and exciting hunting opportunities that are more easily accessible. We suspect grouse hunters will make heavy use of roads and recommend placing greater emphasis on locating girdled patches of trees along (open and seeded) roads to provide plenty of escape cover for grouse and more rewarding hunting experiences. The interspersion of refuge and recreational areas will provide high quality habitat across the entire landscape, but will also minimize disturbance in some portion while providing high quality hunting in the remaining landscape.

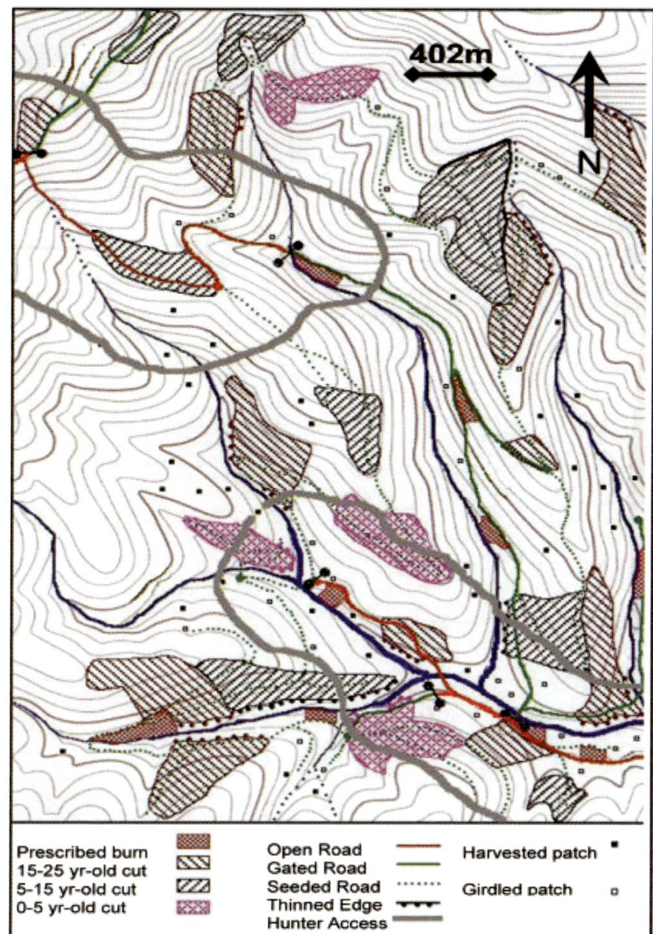

Figure 12.2

Hypothetical 205-ha landscape (adopted from Whitaker 2003) with placement of gates emphasizing "recreational" areas for high quality hunting opportunities over "refuge" areas for ruffed grouse. Recreational areas are ≤ 400 m from an open road and are outlined in gray.

13

Habitat Management

Craig A. Harper, Benjamin C. Jones, Darroch M. Whitaker, Gary W. Norman, Mark A. Banker, and Brian C. Tefft

Appalachian ruffed grouse require a variety of forested habitat as well as openings within the forest. Relative use of habitat by grouse depends on forest type, forest age, and season. Each season brings changes in biological activities of ruffed grouse and the environment in which they live. In the Appalachians, grouse adjust by using forest stands with seasonal foods in or near adequate cover. Optimal habitat affords these resources in close proximity, and they may be found within one forest stand. Gardiner Bump and his crew of researchers in the Catskill Mountains recognized such *interspersion* of cover types and age classes as beneficial during the first in-depth investigation of ruffed grouse (Bump et al. 1947). Subsequent work by Gordon Gullion (1977, 1984a) developed a silvicultural procedure for the Great Lakes States that diversified habitat through rotational harvest of aspen. Following Gullion's recommendation, a patchwork of small clearcuts implemented at 10-year intervals over a 40-year rotation has been shown to support high densities of grouse. In mixed-oak forests where aspen is absent and timber rotations are longer (80–120 years), managers face a more daunting task of providing high-quality cover and diverse food resources over space and time. Although silvicultural practices differ, interspersion of forest age classes and other important habitat features is critical when managing habitat for ruffed grouse in the central and southern Appalachians.

Reproduction, recruitment, and survival determine year-to-year grouse abundance, and positive relationships have been shown between these parameters and high habitat quality (Devers 2005). Lack of nutritious foods and suitable cover are often cited as limiting factors for Appalachian grouse populations (Norman and Kirkpatrick 1984, Servello and Kilpatrick 1987, Servello and Kilpatrick 1988, Long and Edwards 2004a). As a result, habitat manipulation that improves food availability and escape cover can promote population growth (Kubisiak et al. 1980, McCaffery et al. 1996, Stoll et al. 1999, Storm et al. 2003). The *location, proximity, and design* of management units with respect to seasonal habitat requirements in large part determine the success enjoyed by grouse management programs. The task for land managers who want to improve conditions for ruffed grouse is to provide needed habitat in sufficient amount and in an arrangement that makes the area as favorable as possible for grouse.

Forest Management Practices to Improve Ruffed Grouse Habitat

Forests are managed through regeneration methods and various timber stand improvement (TSI) practices (Smith 1986). Sound forest management also involves managing forest roads and openings in an effective and efficient manner (Healy and Nenno 1983). Regeneration methods set back forest succession and allow a new stand to develop. TSI practices, such as thinning, are used to manipulate developing stands and provide additional resources (sunlight and nutrients) to favored tree species and individual trees. To improve habitat conditions for ruffed grouse, the appropriate methods and practices used are determined by site, forest type, tree species composition, stand age, stand history, and the objectives of the landowner/manager. A review of the literature pertaining to habitat management for ruffed grouse reveals numerous recommendations for forest management, with clearcutting ubiquitous in most reports because of the propensity of ruffed grouse to use young forest stands (McCaffery et al. 1996, Stoll et al. 1999, Storm et al. 2003). There are several regeneration methods, but not all are suited for every forest type or situation (Sander et al. 1983, Smith et al. 1983, Tubbs et al. 1983). Careful consideration should be given to the desired stand composition and structure (i.e., vertical and horizontal arrangement of vegetation) before silvicultural techniques are prescribed. Another important factor is the potential impacts of deer herbivory on desired regeneration. Deer population management is a prerequisite to successful forest management in many areas.

Regeneration Methods

Clearcut. Clearcutting is a regeneration method that removes all trees above a specific size from the site, creating an even-aged stand (Smith 1986). This is an efficient technique in terms of harvesting timber, as loggers visit the site only once (over a period of a few weeks, or less, depending on the size of the stand). Clearcutting allows more sunlight to reach the forest floor than other regeneration methods, resulting in vigorous competition among shade-intolerant (e.g., yellow-poplar, black locust, black cherry, pin cherry, and basswood) and other species that sprout and grow rapidly after cutting (e.g., red maple, white ash, and birches) (Beck and Hooper 1986, Lorimer 1992, Elliott and Swank 1994). Less aggressive species (including oaks) that are intermediate in shade-tolerance are often underrepresented in clearcut-regenerated stands, especially on higher-quality sites (Loftis 1990, 1993).

Despite shortcomings of clearcutting for regenerating oaks, clearcut stands provide excellent cover for ruffed grouse, especially 5–20 years after harvest. Grouse may use clearcut stands for escape cover, foraging, nesting, drumming, and brood rearing during this period (Sharp 1963, Scott et al. 1998, Schumacher 2002, Whitaker 2003, Jones 2005, Jones et al. 2008). Beyond 20 years, habitat quality decreases as the canopy closes and grows taller, causing decreases in woody stem density, herbaceous ground cover, and soft-mast production.

Following clearcutting, structural characteristics and species composition of the new stand are largely dependent upon the site. Mesic (moist) sites, such as those found in coves and on north- and east-facing aspects, usually regenerate yellow-poplar, sugar maple, yellow and black birch, black cherry, cucumbertree, basswood, serviceberry, and American beech, with scattered yellow buckeye and northern red oak. Several of these species (birch, cherry, and serviceberry) produce buds that are important winter food for grouse (Servello and Kirkpatrick 1987, Plaugher 1998, Long and Edwards 2004b). Other foods, such as blackberries and blueberries, as well as herbaceous forage, are often abundant following clearcutting.

In mixed-mesophytic and northern hardwood forests, clearcutting regenerates numerous desirable species for ruffed grouse. On drier sites, where oak-hickory forests are more prominent, hard mast (especially acorns) is an important winter food for grouse. Clearcutting oak-hickory stands creates high stem densities desirable for escape cover, however, mast production is eliminated for approximately 40 years (Guyette et al. 2004). Even then, mast production will not equal that of the previous stand if oaks are underrepresented in the regenerating stand. Where advance oak regeneration (regenerating seedlings or sprouts 0.3–1 meter [1–3 ft] tall) is present in the understory, clearcutting may be an effective system for regenerating oak-hickory forests; nonetheless, mast production is still absent for a number of years.

Despite sound forest management research and reasoning, forest management systems have come under extreme scrutiny in recent years by special interest groups. In particular, clearcutting as a regeneration method has come under fire in numerous localities. Forest management options remain numerous on most private, industrial, and state-owned lands; however, forest managers on federal land (National Forests in the Appalachians) are limited in silvicultural options because of litigation surrounding timber management prescriptions. Many private landowners also feel clearcutting is too invasive and thus look for other regeneration methods as aesthetic alternatives to clearcutting. Some alternative methods (that may not be as likely to be appealed by special interest groups) have real potential in promoting regeneration of some important hardwood species.

Shelterwood. The shelterwood method has been used more in recent years for increasing the development of oak regeneration (Loftis 1983, 1990, 1993, Figure 13.1). *This should be a major consideration for land managers in the central and southern Appalachians who are interested in ruffed grouse* (as well as many other wildlife species). Shelterwood harvests occur in two or more stages and produce an even-aged stand (Smith 1986). The initial shelterwood harvest removes a pre-determined amount of the forest canopy, enabling partial sunlight onto the forest floor. This enables existing seedlings of moderate shade tolerance (especially oaks) to better compete with shade-intolerant species and produce advance regeneration. Advance regeneration then is released to grow by subsequent harvest(s) that removes residual overstory (usually 6–8 years post initial harvest).

The amount of overstory retained in the initial harvest depends on desired species composition, the amount of oak regeneration present, site productivity, and regeneration mechanism (seed, sprout, advance reproduction) of both oaks and competing species. There is a fine line in deciding how much overstory to leave to benefit oaks. Too much shade will benefit shade-tolerant species, such as sugar maple, dogwood, and beech, while too little shade will benefit yellow-poplar, red maple, and black cherry.

Where oak-dominated forests are desired and site index (a measure of site potential reflecting the expected height in feet that trees will reach at a target year, usually 50 years for oaks) for oak is relatively low (60–65), the overstory retained may be only 4.5–9.0 m^2/ha [20–40 ft^2/acre] of basal area. On these sites, advance reproduction is usually not a problem as oaks are often the dominant overstory species. Where the site index for oak is high (75–80), more overstory must be retained in the initial shelterwood harvest to suppress shade-intolerant species. In some cases, only overtopped and intermediate trees or just the midstory are removed (or killed with herbicides), which allows relatively little additional light to reach the forest floor (this is termed a *thinning from below*; no dominant or co-dominant trees are removed from the overstory) (Loftis 1990). Where existing oak regeneration is sparse or nearly absent, a thinning from below may be conducted after a good acorn crop to help stimulate germination and seedling establishment. Once advance oak regeneration becomes established (1–1.5 meters [3–5 ft] tall), the overstory may be harvested.

On sub-mesic sites (transition sites between mesic and xeric) where yellow-poplar, red maple, and others are serious competitors, a shelterwood harvest followed by prescribed fire has shown promise (Brose and Van Lear 1998, Brose et al. 1999a, Brose et al. 1999b). Three to five years after the initial shelterwood harvest, a growing-season

Figure 13.1

Shelterwood regeneration cuts.

Top: One year post-harvest provides open cover with potential for brood habitat. Note that the skid trail was constructed through the middle of the harvest unit.

Bottom: Seven years post-harvest, a shelterwood cut has developed into high-quality grouse cover. The planted skid trail provides travel corridors.

Photos: Craig H. Harper.

fire is used to top-kill all trees in the stand. Young oaks arising from an existing root system are then able to send up a vigorous stem the year following fire and compete with other species. On more xeric sites, especially south- and west-facing slopes and ridge tops, establishment of oak regeneration is less difficult. Several species of oaks (including white, chestnut, black, and scarlet) reproduce vigorously on drier sites following harvest.

Initial shelterwood harvests may leave as little as 10–30% of the original canopy cover. This results in regenerating stem densities and species composition similar to that following a clearcut. Regardless of the amount of residual overstory left standing, **it is critical that high-quality mast producing trees (especially oaks) are retained instead of other species with less value to ruffed grouse**. A good mixture of oaks (species from both white and red oak groups) should be retained to offset variation in mast production by different species.

Shelterwood harvests can benefit grouse in several ways. Depending on the site, opening the forest canopy increases groundcover and enhances foraging and brooding opportunities. A greater herbaceous response can be expected on mesic sites, while a greater woody response can be expected on xeric sites. Soft-mast production (e.g., blackberry, raspberry, blueberry, huckleberry, and pokeberry) also can be expected to increase 2–5 years post harvest, increasing both food availability and high-quality brood cover (Greenberg et al. 2007). Escape cover is enhanced as midstory stem density increases following harvest. The benefits of shelterwood harvests over clearcutting are the retention of mature, mast-producing oak while advance regeneration is developing, provision for oak in the future stand, and retention of mature trees for aesthetic purposes. Acorns are a nutritious food that can influence survival and recruitment of Appalachian ruffed grouse. Therefore, stands that intersperse mature oaks with woody sapling cover will benefit grouse in the region. In North Carolina, our radio-tagged grouse began using stands harvested by the shelterwood method three years after initial harvest, prior to removal of residual canopy trees (Jones and Harper 2008; Figure 13.1).

Another advantage of the shelterwood method is that loggers have to come back into the stand one or more times over several years after the initial harvest and remove the residual overstory. Although this is less efficient in terms of harvesting timber, it is beneficial to grouse because another flush of herbaceous cover and soft-mast production can be expected after each harvest. This benefit is reduced, however, if invasive non-native plants (e.g., japangrass) are allowed to pioneer into those disturbed areas.

Two-aged System A two-aged system represents a planned sequence of treatments designed to regenerate and maintain a stand with two age classes where select "reserve" trees are retained after the initial harvest to attain goals other than regeneration (Figure 13.2). Reserve trees not only increase the future value of the stand, but also can provide wildlife benefit and make the stand more aesthetically pleasing after harvest (Smith et al. 1989).

A shelterwood with reserves (or irregular shelterwood) produces a stand of two distinct age classes—a residual mature overstory with developing regeneration below. The difference between a shelterwood and an irregular shelterwood is that the regeneration period is extended with an irregular shelterwood, resulting in a new stand that is not really even-aged, but two-aged. The stand will include two age classes for at least 20–30% of the rotation and often for the entire rotation, depending upon objectives. Normally, 3.4–5.7 m^2/ha (15–25 ft^2/acre) in dominant, co-dominant, and good intermediate crown-class trees are retained; however, a higher residual basal area may be retained if desired. As with a shelterwood, regenerating stem density is greater when less overstory is retained. Trees retained in an irregular shelterwood are chosen based on their capacity to produce seed and increase in value until the regenerating stand is harvested. When few oak seedlings are present, a thinning from below following a good mast crop can be used to help stimulate and increase oak regeneration before harvest.

Another two-aged regeneration method is a clearcut with reserves. This method is similar to an irregular shelterwood except a clearcut with reserves retains no more than 1.1–2.2 m^2/ha (5–10

Figure 13.2

A two-aged stand 10 years after harvest. The high-quality white oak that was retained provides hard mast within the cover supplied by the regenerating stems. *Photo: Craig H. Harper*

ft^2/acre) of basal area post harvest and there is no plan to harvest the overstory until the end of the rotation of the regenerating stand.

Regeneration methods that produce two-aged stands show great promise in creating optimal habitat for Appalachian grouse; however, *it is imperative that oaks with good growth form and mast production potential are retained as residuals*. When high-quality oaks are selected, an irregular shelterwood that retains the residual overstory for at least 30–40 years (until the regenerating stand begins to produce mast) is the best regeneration method to improve habitat for ruffed grouse when harvesting oak-hickory stands in the Appalachians. Our research indicated a strong inverse relationship between grouse home-range size and mast crops in oak-hickory stands (Whitaker 2003). When stands are clearcut, there is a time lag in hard-mast production while trees mature (at least 30–40 years). During that period, grouse must balance time spent in young regenerating stands and time spent foraging among mature oaks. Two-age stands provide both food and cover, allowing grouse to forage on acorns and other foods without increasing risk of predation. In West Virginia, flowering dogwood, serviceberry, and pin cherry were present in two-aged stands, and grapevines occurred in 58% of the co-dominant reproduction stems (Miller and Schuler 1995). Similar to shelterwoods, grouse also began using irregular shelterwoods in North Carolina three years after harvest (Jones and Harper 2008).

Group Selection. The group-selection method mimics small-scale canopy gaps created by low-intensity natural disturbance events. Group selection harvests small groups of trees within a stand over time, creating a mosaic of even-aged patches within an uneven-aged stand (Smith 1986). By using group selection harvests, a percentage of young forest cover with high stem densities can be maintained across the stand while avoiding visual impacts of larger even-aged harvests. The size of group selection harvests ranges from a small area occupied by a few trees (0.04 hectare [0.1 acre]) to nearly 0.8 hectare (2 acres).

Size of group selection cuts may influence stand composition and structure (Dale et al. 1995). Although site quality, moisture regime, and previous stand composition are the primary influences on future stand composition, larger group harvest units (over 0.4 hectare [1 acre]) are more likely to result in shade-intolerant species, such as yellow-poplar and basswood. Shade-tolerant (sugar maple, beech) and intermediate species (oaks, birches) may be more prevalent in smaller group harvests. In North Carolina, yellow-poplar, sweet birch, and red maple sprouts dominated regeneration within small group openings (less than 0.08 hectare [0.2 acre]) on mesic sites, while oak regeneration was plentiful as a result of diffuse sunlight on the forest floor around the periphery of each patch. As with even-aged methods, the presence of advance oak regeneration is an important consideration before implementing group selection harvests in oak-hickory stands.

The optimal density of group selection harvests in a given stand is debatable. If the character of a mature stand is desired, the density of cuts should be low. If visual impact is not an issue, the density of group cuts can be increased. Positioning one group selection cut per 4 hectares (10 acres) would place patches approximately 240 meters (800 ft) apart, harvesting 2.5–6.25% of the stand. Thus, grouse would be able to remain within about 120 meters (400 ft) of escape cover when foraging in an adjacent mature stand.

Although not documented or demonstrated, concern has been expressed that the group-selection method creates isolated pockets of habitat. To relieve this concern, thinning between groups would minimize edge effects, increase understory stem density, and improve groundcover conditions and connectivity between groups. Regardless, group cuts should be well interspersed to increase cover and foraging opportunities for ruffed grouse in mature stands. Groups themselves also may serve as stepping stones, and thus act as travel corridors. The group-selection method should not be viewed as a substitute for even-aged management, but rather as a complement, serving to connect young forest stands and improve conditions for grouse over a broader area.

Grouse broods often use small canopy gaps and edge habitat within otherwise mature forest cover (Stewart 1956, Thompson et al. 1987). In North Carolina, brooding hens used edges of group cuts four years after harvest (Jones and Harper 2008; Fig 13.2). These cut units contained abundant groundcover and were located within 80+-year-old mixed-oak stands—an important forest type for broods on the study area. Evidence that group-selection harvest units enhanced brood habitat was provided by the observation that broods using mixed oak stands lacking group cuts were often associated with canopy gaps, which were similar in composition and structure to the group-selection harvest units.

Timber Stand Improvement (TSI) Practices

Thinning and Wildlife Retention Cuts. Hardwood stands can be thinned prior to maturity to influence stand species composition and increase sunlight and nutrients to promote growth and development of selected residual trees. Growth and yield are increased most if the stand is first thinned at about age 20 and continued at about 10-year intervals until age 60–70, though timing will depend on species composition and site quality (Smith 1986). Thinning has real implications in ruffed grouse management if those tree species that do not produce preferred food resources (e.g., maples, yellow-poplar, ashes, and sourwood) are targeted for removal, while more desirable species (e.g., oaks, black cherry, serviceberry, birches, American beech) are retained (Figure 13.3). Thinning undesirable trees also allows increased sunlight into the stand, stimulating understory development. As with regeneration methods, understory composition following treatment will depend on the site. Typically, mesic sites will produce more herbaceous vegetation, while xeric sites will produce more woody cover (Jackson et al. 2006). Regardless of site, soft-mast production by species such as blueberry, huckleberry, blackberry, and raspberry can be expected to increase 2–5 years post treatment. In transitional and xeric mixed hardwoods, soft- and hard-mast producing species favored by grouse (e.g., oaks, serviceberry, and blackgum) are retained in the overstory, while others are targeted for removal. In mesic stands where oaks are less prominent, retention of black and pin cherry, birch, American

Figure 13.3

Typical thinned stand, showing the increased understory growth that provides cover important to grouse.
Photo: Craig H. Harper

beech, and serviceberry and release of herbaceous understory for foraging opportunities and high-quality brood habitat are the primary objectives of thinning to improve grouse habitat.

By definition, thinning is conducted in immature stands only. Thinning past age 60–70 does little to increase the growth of residual trees; thus, thinning operations are not warranted economically to increase timber production in mature stands. Mature stands can be enhanced for grouse and other wildlife through a *wildlife retention cut* designed for wildlife objectives (Jackson 2002, Basinger 2003, Gordon 2005). The objective of a wildlife retention cut as a TSI practice is to reduce percent canopy closure to approximately 60–70% (or less, if desired, for increased stem density) by killing selected trees with herbicide injection or girdling and spraying the wound with herbicide. Targeted trees are treated and left standing as snags; they are not felled or removed from the site, but allowed to fall apart over time providing coarse woody debris. Trees are selected based on mast-producing capability and overall form. Non-mast producers and trees with poor growth form are targeted first to reduce canopy closure to the desired level. Mid-story soft-mast producers (e.g., dogwoods) are normally retained unless they are so abundant they cast an inordinate amount of shade.

Wildlife retention cuts not only stimulate understory development, but also enable crowns of residual trees to develop more fully when adjacent trees are killed. Mast is produced near the ends of twigs within a tree's crown. As a crown increases in diameter, more twig ends are present to produce additional mast. By default, the larger the crown, the more potential the tree has to produce fruit. Thus, it is possible for a stand to produce *more* mast with *fewer* trees *while* supporting enhanced understory cover. A wildlife retention cut is not the same as *crop tree release*. Crop tree release is conducted only in immature stands by removing competing trees to enhance the growth of the desired species; stand-wide reduction in canopy closure is not an objective. In addition, a wildlife retention cut is not similar to a *diameter-limit cut*, which removes all trees above a predetermined diameter and gives no consideration to species composition. Diameter-limit cuts normally amount to "high-grading," (i.e., taking the best or "highest grade" trees and leaving the rest) and are not recommended with regard to forest ecology or ruffed grouse management.

Salvage Operations. Forest succession in the Appalachians is commonly driven by wind, ice, insects/disease, and fire. Salvage harvest operations are often feasible after stand disturbance, particularly wind, ice, and insect/disease. This offers an opportunity to make good use of merchantable timber that might not be harvested otherwise, and improves habitat conditions for ruffed grouse at the same time. Depending on the source and level of damage, salvage cuts often resemble shelterwood or two-age harvests as there are usually residual trees remaining after the operation (Figure 13.4). In 1995, Hurricane Opal caused extensive blowdown of forest stands in the southern Appalachians. Following salvage operations, researchers at Coweeta Long-Term Ecological Research Station in North Carolina measured greater understory plant diversity in salvage areas compared to undamaged stands and recent clearcuts. The greater understory diversity was largely a result of shading provided by residual trees and slash, as well as pit-and-mound topography (soil disturbance) created by uprooted trees (Elliott et al. 2002).

The opportunity for salvage operations to improve ruffed grouse habitat in the central and southern Appalachians should not be overlooked. A major consideration when implementing a salvage operation should be to address the composition and quality of residual trees. Poor-quality, previously suppressed stems should be felled during salvage operations to help ensure future stand quality. Likewise, non-favored species for

Figure 13.4

Salvage cut in North Carolina seven years after harvest. It is clear the structure is suitable for grouse, and similar to that resulting from clear cuts or shelterwoods. *Photo: Craig H. Harper*

ruffed grouse also should be killed or felled to positively influence future stand composition.

Prescribed Fire. Although once commonly used, fire has been suppressed in the Appalachian region for nearly 100 years, altering many of the associated forest types and wildlife communities (Van Lear and Waldrop 1989, Johnson and Hale 2002, Van Lear and Harlow 2002). Fortunately, forest and wildlife managers are realizing the positive benefits of fire and using it more often in the Appalachians, especially to reduce fuels and foster oak regeneration. This has proven most beneficial for ruffed grouse and wild turkeys, especially in oak-hickory forests where controlled burning can enhance understory structure important for winter foraging and brooding habitat (Rogers and Samuel 1984, Pack et al. 1988, Jackson et al. 2006).

In North Carolina, fire was prescribed in an upland oak forest during March 2002. By 2004, the treated area (approximately 285 hectares [700 acres]) supported a diverse herbaceous community, which was used almost exclusively by several grouse broods. Midstory conditions also were improved by sprouting flame azalea, buffalonut, and mountain holly. In western Virginia, we documented positive results in young hardwood clearcuts following prescribed fire, including increases in invertebrate abundance and soft mast-producing plants (Whitaker et al. 2004). Grouse broods in the Appalachians select areas with abundant herbaceous vegetation, especially forb and fern cover, but also low-growing woody cover, such as blueberries and huckleberries (Scott et al. 1998, Haulton 1999, Fettinger 2002, Jones et al. 2008).

Prescribed fire in the Appalachians is restricted primarily to oak-hickory forests and other forest types associated with southern and western exposures and ridge tops (Van Lear and Waldrop 1989). This offers numerous opportunities for habitat enhancement, especially where oak-hickory forests comprise 50% or more of the available forest cover. When burning oak-hickory stands, fire often feathers into coves and more mesic forests types, but intensity is much less and these areas rarely burn. In fact, when burning relatively large areas (80–200 hectares [200–500 acres]; which is usually necessary on national forests where there is a lack of roads or firebreaks), coves, creeks, and northern/eastern exposures are commonly used as natural firebreaks. This provides an exceptional mosaic of conditions across the burned area, which is most favorable for ruffed grouse.

Fire intensity is determined by fuel load and moisture content, wind, humidity, temperature, and atmospheric conditions (Wright and Bailey 1982). Land managers must balance fire intensity with existing site conditions to create the desired habitat structure and composition. For example, a relatively cool fire may be used to consume the litter layer and promote an herbaceous understory, while a hot fire is necessary to reduce extensive coverage of mountain laurel and allow adequate light to the forest floor to stimulate the seedbank. Depending on stand age, stocking, and percent canopy cover, either thinning or a wildlife retention cut is sometimes desirable prior to burning. Basal area will fluctuate among sites, but reducing canopy closure to 60–70% normally allows sufficient sunlight into the forest floor to develop the desired understory structure for brood habitat and will also promote additional soft-mast production (Jackson et al. 2006).

The historical occurrence of fire in the Appalachian region has been debated, but historical dendrochronological evidence shows lightning- and Native American-ignited fires occurred every 3–25 years in those stands that would burn, depending on the site and climatic conditions (Frost 1998). As related to habitat management for grouse, understory composition and structure, midstory characteristics, fuel load, and site determine fire frequency. On drier sites, it is common for woody species to dominate the understory, while more mesic sites have greater herbaceous cover. This can influence fire return interval. More frequent fire (annually to every 2 years) on drier sites can be used to stimulate increased herbaceous cover.

Special Considerations

In addition to forest regeneration and TSI practices, there are other activities that can improve ruffed grouse habitat, often dramatically, albeit on a smaller scale. Spring seeps occur where groundwater percolates to the surface and forms a saturated area. Even during periods of deep snow, these areas are often snow-free (Healy 1977, Healy and Pack 1983, Wunz et al. 1983). Reducing canopy cover to approximately 50% around spring seeps allows increased light into the site and can promote herbaceous groundcover and shrub growth, which produces many fruits and seeds that are eaten by grouse in mid-winter. Hard-and soft-mast producers should be retained. Where soft-mast producing trees and shrubs do not exist, they can be planted around seeps when adequate sunlight is available through thinning. Shrub species such as hawthorn and crabapple have been successfully established around spring seeps after thinning. When stands containing seeps are regenerated, trees surrounding the seep (0.1–0.4 hectare [0.25–1 acre], depending on the site) should be left uncut to provide mast and perches for grouse.

On many areas, old homeplaces are present. These sites usually support relatively high stem densities and often hold grouse, especially if old fruit trees remain. Thinning around existing fruit trees to allow adequate light, pruning excess limbs, and fertilizing stimulate increased growth and fruit production. Planting additional soft-mast producers on these sites and maintaining them as wildlife orchards benefit ruffed grouse and many other species. Tree/shrub species that should be considered include apples, crabapples, pears, plums, hawthorn, persimmon, mulberry, serviceberry, elderberry, and Carolina buckthorn.

Grapes are an important food of grouse in the Appalachians. Grapevines not only should be retained, but promoted when possible. Grapevines are often found on mesic sites, often in narrow coves just above or below logging roads. These sites can be improved by thinning undesirable trees and allowing sunlight into the site to stimulate additional groundcover and stem density and improve conditions for foraging grouse. Teepee-style grape arbors can be created by felling adjacent, undesirable trees against the tree supporting the main vine. Grapevine growth is not necessarily desirable to foresters and loggers as they can suppress valuable trees and make timber harvest more difficult. Where habitat management for ruffed grouse is an objective and grapevine growth is so excessive it poses a danger, then the tree(s) should be left standing. If the tree(s) supporting grapevines is not desirable for grouse, then it can be killed and left standing as in a wildlife retention cut.

Although grouse may roost in pines, there is not a strong selection for roosting in pines in the Appalachians (Whitaker 2003). As a result, there is no need for planting pines as a means to improve roosting cover in this region. In fact, isolated evergreens (e.g., large hemlocks) can serve as predator traps for grouse as hawks and owls easily develop a search image and concentrate on these trees, especially when individuals are retained in regenerated stands. Gullion (1990) felt any advantage (e.g., foraging sites or protected roost sites) grouse gained by using coniferous cover was offset by a higher risk of predation and shorter survival (Gullion and Marshall 1968, Gullion 1981a). If suitable roosting cover is limited on a particular area, a more sound recommendation is to simulate blowdowns by felling small groups of trees (similar to group selection harvests) because grouse roosting in mature stands selected microsites having locally high stem densities (Whitaker 2003).

Forest Roads and Openings

Forest roads, such as old logging roads, and managed openings provide critical habitat for ruffed grouse in the Appalachians (Whitaker 2003, Jones 2005). Grouse (male and female, adult and juvenile) select forest roads during various seasons throughout the region. Forest roads and openings are important foraging areas, especially within oak-hickory-dominated forests during years with little mast production. In most areas where grouse are found in the Appalachians (especially National Forest land), forest roads and openings compose less than 1% of the land cover. Because they are such a critical habitat, managing roads and openings in an effective and efficient manner is paramount to ruffed grouse management.

The Importance of Forest Roads for Ruffed Grouse. When seeded and managed properly, forest roads can be turned into "linear openings" for wildlife, providing increased habitat interspersion, much-needed high-quality forage, and attractive brood habitat. Forest roads are an important habitat feature for ruffed grouse in the Appalachians (Schumacher 2002, Whitaker 2003, Jones et al. 2008). More than 90% of the 326 grouse crops collected from forest roads as part of the ACGRP during March of 2000–2002 contained herbaceous leaves and flowers (Long and Edwards 2004b). These foods represented 25% of all material in the crops over the three-year period. The vast majority of the herbaceous leaves eaten by grouse were from clover, cinquefoil, wild strawberry, avens, hawkweed, and birdsfoot trefoil. Coltsfoot was the most frequently eaten flower. Interestingly, though orchardgrass was the predominant cover on most of the forest roads where grouse were collected, no orchardgrass was found in any of the grouse crops. In fact, of 326 crops examined from six states, **no grass of any kind was found in measurable amounts.** From this research, it is apparent forest roads dominated by legumes and other forbs are most beneficial to ruffed grouse.

Considerations With Sedimentation, Road Construction, and Road Closure. The primary consideration in managing forest roads is preventing erosion and sediment run-off. Openings are not as prone to wash and are not typically located on steep slopes. Proper road construction is the best deterrent to prevent erosion and siltation into streams (Swift 1984, 1985). Long-term research at the Coweeta Hydrologic Lab showed improper road construction leads to more than 95% of the siltation into streams following logging operations, not the logging operation itself (Swift 1988). Logging roads should be constructed (or repaired) following Best Management Practices set by state forestry agencies. Steep slopes should be avoided. Water bars and broad-based dips should be created where water flow and drainage may be a problem. Roads must be seeded to establish vegetation as quickly as possible after the logging operation is finished to prevent erosion. Finally, all logging roads should be gated to reduce vehicular travel, which damages established vegetation and may lead to undue pressure on the grouse population during the hunting season. Roads (whether gated or not) that receive considerable traffic should be graveled. It is unrealistic to expect vegetated roads to sustain regular vehicular traffic. This inevitably creates "two-tracks"—tire lanes worn down to mineral soil that channel water flow during heavy rain events and lead to increased siltation.

"Daylighting" Roads. The foremost consideration when seeding forest roads is available sunlight. To achieve adequate germination and growth after seeding, the road should receive at least four hours of direct sunlight per day. This may not be a problem for roads recently created, but the canopy of adjacent trees will slowly shade the road over time. Unless the adjacent stand has been thinned or regenerated recently, roads require "daylighting"—that is, removing select trees along at least one side of the road (at least nine meters [30 ft] on one or both sides of the road) to allow sufficient sunlight for herbaceous cover to establish and grow in the road. This practice alone improves conditions for grouse, as herbaceous groundcover and woody stem density is increased along one or both sides of the road, providing a soft edge into the adjacent forest. Not all trees have to be removed. Again, species that provide little benefit for grouse can be removed, while scattered beneficial trees are retained.

Liming and Fertilizing Roads. Most of the soils in the Appalachian region are acidic (pH <5.8) and require liming to increase pH. Most plantings on forest roads benefit from a pH closer to neutral (6.0–7.0) because soil nutrients are made more available through chemical reaction and organic decomposition, and soil bacterial activity is increased, which influences nitrogen-fixing bacteria and organic matter decomposition (Donahue et al. 1983, Ball et al. 2002). Depending on soil characteristics, two tons (or more) of lime per acre are often required to increase pH for optimum plant growth in the Appalachian region. Only by collecting soil samples and having them tested is it possible to know how much lime (or fertilizer) is actually needed. It is important to realize the full effect of liming on soil pH is not realized until approximately six months after application. Soil pH will slowly decline to original levels 5–10 years after liming unless the site is top-dressed with additional lime as recommended from a soil test. Phosphate (P_2O_5) and potash (K_2O) levels often test low (<20 kg P_2O_5/ha [18 lbs P_2O_5/acre]; < 100 kg K_2O/ha [90 lbs K_2O/acre]). To expect most planted materials to produce at least 75% of their potential, 21–32 kg/ha (19–30 lbs/acre) of P_2O_5 and 100–180 kg/ha (91–160 lbs/acre) of K_2O should be available. Although commonly applied, balanced fertilizers (e.g., 15-15-15) are rarely needed, especially if legumes (e.g., clovers) are planted. Symbiotic bacteria produce nitrogen within nodules attached to the roots of legumes. As a result, other fertilizers, such as triple super phosphate (0-46-0) and muriate of potash (0-0-60), can be used to increase P_2O_5 and K_2O levels without unnecessarily increasing nitrogen levels.

Considerations for Planting Forest Roads. If roads are to provide nutritional benefit for grouse, plants that are nutritious and actually eaten by grouse must be established (Figure 13.5). To provide attractive brood habitat for grouse, plants that restrict travel and feeding by chicks should not be planted. At the same time, consideration must be given to preventing erosion and sediment flow. Plants that germinate and establish root systems quickly are needed to hold the soil together and prevent washing.

Plants that best meet these requirements are wheat and clovers (Harper 2008). One mixture that has worked well in the Appalachian region

Figure 13.5

Gated roads provide high quality grouse cover.

Top: Brambles supply cover and soft mast, while the herbaceous cover in the road center will be used by broods.

Bottom: Ladino clover was planted to provide forage and cover, and the sides of the road were sprayed with a selective herbicide to control encroaching japangrass.

Photos: Craig H. Harper

includes (per acre, 0.4 hectare) 22.7 kilograms (50 lbs) of wheat and 3.6 kilograms (8 lbs) of ladino white clover. For best results (if seed are top-sown), wheat should be sown and lightly disked prior to sowing the small clover and trefoil seed. After planting, the seedbed should be firmed using a cultipacker. (Note: Cultipacking after seeding greatly improves germination rates and initial growth.) If drilled, the wheat and clover seed must be planted in separate seed boxes. All legume seed should be inoculated with species-specific inoculant unless pre-inoculated seed are sown. Ladino and white Dutch clover require *Rhizobium leguminosarum* biovar *trifolii* (Harper 2008).

Perennial cool-season grasses (e.g., tall fescue, orchardgrass, bromegrasses, bluegrass, and timothy) are not recommended in seeding mixtures because they are slow to establish (as opposed to annual grasses), do not produce forage or seed that are eaten by grouse, produce a dense structure at ground level (precluding travel by chicks), do not support high invertebrate populations (as compared to forbs), and out-compete favored plants (such as clovers) within two growing seasons, leaving a strip of rank grass that provides little, if any, benefit to ruffed grouse (Harper et al. 2001, Fettinger et al. 2002, Jones 2005, Harper 2008). Many land managers have been led to the false assumption that it is necessary to include perennial cool-season grasses (especially orchardgrass or tall fescue) in a mixture sown on forest roads. **This is not true and certainly counterproductive for wildlife!** (Figure 13.6)

Figure 13.6

Perennial cool-season grasses, such as this orchardgrass, should not be included in a mixture when planting forest openings or roads. The dense nature of these grasses at ground level makes travel by grouse broods exceedingly difficult, and outcompetes desirable forbs, such as clovers, cinquefoil, and wild strawberry, which are important grouse foods. *Photo: Craig H. Harper*

The Value of Annual Grains Over Perennial Grasses. Annual cool-season grains can be planted in spring or fall, germinate within four days of rainfall, and establish a root system capable of preventing erosion within a few weeks, depending on local conditions. Perennial grasses (cool- or warm-season) require much longer to germinate and become established. Annual cool-season grains (especially wheat) also produce seed that may be eaten by grouse and other birds, and serve as a nurse crop while clovers become established underneath. By the time the annual grain completes its life cycle, clovers and other forbs have become established and cover the site.

Renovating Species Composition Along Forest Roads. Plant composition along forest roads can be improved with herbicides and (if desired or needed) top-sowing or drilling seed. Roads dominated with perennial cool-season grasses can be renovated by spraying a glyphosate herbicide over actively growing grass in late September. Approximately one month prior to spraying, the road should be mowed to reduce senescent stems and encourage fresh grass growth. Also at that time, lime and fertilizer should be applied as recommended by a soil test. When the grass is approximately 15–25 centimeters (6–10 inches) high, it should be sprayed. Herbicide label recommendations should be followed. Approximately two weeks after spraying, clovers and trefoil can be top-sown over the dying/dead grass. As the thatch begins to decay, the planted seed will germinate. Instead of top sowing, seed can be drilled. The existing root systems hold the topsoil intact through winter and by March the road will be lush and green with high-quality forage for ruffed grouse and other wildlife. If unwanted residual grass begins to re-appear, it can be selectively removed using a grass-selective herbicide.

Japangrass is a non-native annual, warm-season grass that has become problematic in many areas, especially along roadsides. Japangrass often dominates these sites, inhibiting growth of native plants and suppressing the seedbank, similar to tall fescue and other non-native perennial cool-season grasses. Japangrass can create poor structure for grouse with broods and decrease food availability. Fortunately, this invasive grass can be removed fairly easily by spraying selective herbicides in mid to late summer (before producing seed), leaving various forbs that may provide attractive brood cover and/or food. Repeat applications are necessary to eradicate residual seed from the seedbank.

Site Considerations. Ladino clover is a cool-season legume that does best on moist or well-drained sites that are not too dry (Ball et al. 2002). Ladino clover typically does not persist long on south- and west-facing slopes, especially during hot, dry summers, unless at relatively high elevations (>4,000 feet) or northern latitudes where cooler temperatures may prevail during summer. Under harsh, dry conditions, ladino clover wilts down and the cover generally thins. This is not a problem for grouse (fortunately for land managers) because many forbs are available in the seedbank (that collection of seed occurring naturally in the top few inches of soil) waiting to germinate. Many of these forbs, such as wild strawberry, cinquefoil, avens, hawkweeds, and ragwort, are readily eaten by ruffed grouse. Naturally occurring forbs also harbor higher invertebrate populations than grasses and provide attractive brood habitat that allows travel under the protection of forb cover (Harper et al. 2001). These are the very reasons female grouse with broods use forest roads so often during the early brood-rearing season!

Land managers should use these site limitations to their advantage. Roads on exposed south- and west-facing slopes should be managed as brood habitat—that is, allowed to re-vegetate to naturally occurring forbs and grasses. Native forbs germinating from the seedbank are adapted to local soils, and soil amendments such as liming and fertilization are often not needed to improve brood habitat. In addition, these roads do not need to be mowed every year. Mowing every other year is sufficient. Once cool-season grasses have been eradicated, a deep thatch and dense structure at ground level no longer persist and travel conditions for chicks remain open.

Those roads located on north- and east-facing slopes and those in riparian areas and toe-slopes should be planted to forages that will provide needed nutrition in fall through spring. These roads should be top-dressed with lime and fertilizer as appropriate. Mowing is needed only to reduce weed competition. For best results, weeds should be suppressed using selective herbicides. If undesirable grasses become problematic, selective herbicides can be used to kill the grass and not harm the legumes and other forbs. Refer to Harper et al. (2007) and Harper (2008) for specific herbicide recommendations and other management strategies.

The Importance of Early Successional Forest Openings. Forest openings are used by many wildlife species. As with grouse, a major limitation for white-tailed deer in the Appalachians is nutrition, especially during fall/winter months when there is a poor mast crop. Bears frequent forest openings to forage on clovers and soft mast (e.g., blackberries, blueberries). Forest openings provide critical habitat for wild turkeys, raptors (e.g., red-tailed hawks, red-shouldered hawks, great-horned owls, and saw-whet owls), and songbirds (e.g., indigo buntings, common yellowthroats, eastern towhees, dark-eyed juncos, brown thrashers, gray catbirds, and chestnut-sided warblers). Several species of small mammals (e.g., rabbits, groundhogs, meadow voles, white-footed mice, big brown bats, hoary bats, and red bats) are strongly associated with forest openings. Openings can be managed in a way that is compatible for all of these species, including ruffed grouse, if attention is given to the site and the specific habitat needs of those species.

The Importance of Forest Openings for Ruffed Grouse. The ruffed grouse is a bird of the forest. Appalachian grouse use forest openings, but not to the extent they use forest roads. In North Carolina, adult grouse (without broods) did not use openings (Jones 2005). Brooding females used

WHAT ABOUT WARM-SEASON GRASSES?

Native warm-season grasses (NWSG) have been promoted in recent years to improve habitat for small game (namely bobwhite quail and rabbits) and grassland songbirds (e.g., grasshopper and Henslow's sparrows, eastern meadowlarks, and dickcissels) (Heard et al. 2000, Washburn et al. 2000, Dimmick et al. 2002, Giuliano and Daves 2002, Dykes 2005). NWSG (especially big, little, and broomsedge bluestem, indiangrass, switchgrass, and sideoats grama) are recommended for these species because of the cover and structure they provide (Harper et al. 2007). NWSG are bunchgrasses and, when sown and managed correctly, contain open ground between bunches, which allows small wildlife to travel through the field and allows forbs to germinate and grow amongst the grasses. NWSG are rarely, however, used as forage by wildlife. Their value for wildlife is in the cover they provide.

NWSG have been recommended by some forest managers in the Appalachians for planting forest openings and along forest roads, particularly within National Forests. Strong consideration should be given to the existing seedbank before attempting to establish NWSG (Dickerson et al. no date, Harper et al. 2008). On most sites, the seedbank in forest openings contain several native warm-season and native cool-season grasses (NCSG). Broomsedge bluestem, little bluestem, purpletop, poverty oatgrass, beaked panicum, deertongue, and Canada wildrye already exist on many sites. More importantly, the seedbank almost always contains a rich diversity of native forbs and brambles (e.g., blackberry, raspberry, partridge pea, old-field aster, black-eyed Susans, firepinks, bluets, fleabanes, mints, tick trefoils, lyre-leaved sage, robin's plantain, goldenrods, etc.), which is the primary consideration for early brood-rearing cover and invertebrate availability. Generally, forbs provide better foraging and brood-rearing cover than grasses.

If native grasses are sown in wildlife openings, a low seeding rate (mixtures should not exceed 1.8 kg [4 pounds] Pure Live Seed per 0.4 ha [acre]) should be used to retain the integrity of an open structure at ground level, allowing travel and foraging by grouse broods and promoting germination and growth of forb cover. Native grasses that might be considered for planting in forest openings within the central and southern Appalachians include (varieties in parentheses) little bluestem (*Aldous*), big bluestem (*Niagra*), sideoats grama (*El Reno*), deertongue (*Tioga*), Canada wildrye (*Mandan*), and Virginia wildrye. A wide variety of forbs may (and should) be added to the grass mixture, including New England aster, butterfly milkweed, partridge pea, lanceleaved coreopsis, purple coneflower, *Heliopsis* sunflowers, roundhead lespedeza, wild bergamot, evening primrose, mint, and others. Species planted should be determined by site conditions.

NWSG are not recommended for planting on forest roads, for several reasons. The biggest limitation for Appalachian ruffed grouse is nutrition, particularly during fall/winter. Not only are NWSG dormant during the fall/winter, our research showed grouse do not eat grasses during the pre-breeding period. Secondly, escape cover is not needed *in* the road, it is already available *alongside* the road, as slash, brambles, and brush are plentiful. High-quality forage (e.g., clovers/birdsfoot trefoil, and various naturally occurring forbs) and associated invertebrate populations are the most important considerations. When given the choice, it is be preferable to plant forb mixtures rather than grasses.

— Craig A. Harper

openings, but only around the periphery. Reasons for this are not entirely clear. Most openings contained considerable orchardgrass cover—an obvious deterrent to brood travel. Along the edge of openings, however, were bramble growth, scattered slash, and various forbs—structure quite similar to that found within group selection cuts and natural canopy gaps, which were preferred brood habitat. Distance to edge may be the key. Regardless of the structure within field interiors, grouse with broods simply may not feel comfortable venturing far from the protective cover of the wood's edge. For these reasons, we believe smaller (less than 0.8 hectare or 2 acres) and irregularly shaped openings are best for ruffed grouse in the Appalachians.

The first step in making forest openings attractive to ruffed grouse and many other species that use forest openings is to eradicate the non-native, perennial cool-season grasses (e.g., orchardgrass, tall fescue, bromegrasses, timothy, and bluegrass). This can be accomplished only with the use of herbicides. Plowing and/or disking will not eradicate these grasses. They **will** return in later months. Burning, without the use of herbicides, may only increase their vigor.

Openings must be evaluated for site limitations to determine appropriate management options. Attention should be given to the native (or naturalized) plant community that will arise from the seedbank. Once the competing sod-forming grasses are removed from the site, the seedbank is able to respond, often resulting in an amazing diversity of forbs, grasses, and ferns, which creates optimal brood habitat for grouse and turkeys, nutritious summer forage for deer, rabbits, and groundhogs, and usable nesting cover for indigo buntings and common yellowthroats.

"Natural" openings are best managed with prescribed fire or disking, **not** with a bushhog (mower)! Fire and disking recycles nutrients into the topsoil, stimulates germination from the seedbank, creates optimum structure at the ground level for foraging and movement, and increases seed and invertebrate availability. Mowing should be avoided, especially during the nesting/brooding season (May–August), because it destroys critical brood (for grouse and turkeys) and nest (for songbirds) cover during the time it is needed most. And, as anyone who has ever mowed an opening in the summer knows, mowing also kills wildlife directly. Fawns, bird nests with hatchlings, and young rabbits are commonly killed by rotary mowers during the summer months. Mowing also creates less-than-desirable conditions on the ground for grouse and turkey broods as mowed debris accumulates on the surface, making it much more difficult for chicks to move through the field and limiting the ability of broods and other birds to glean seed from the ground because seeds are buried under a deep thatch layer. In addition, accumulating debris from mowing inhibits the seedbank from germinating, which leads to decreased plant diversity. If prescribed fire and disking are absolutely not possible, mowing should be delayed until late winter, providing winter cover within the opening as long as possible.

High-quality fall/winter forage can be provided by planting the wheat/clover/birdsfoot trefoil mixture recommended previously. Other forages that might be considered in openings include chicory (a forage variety, such as *Puna* or *Oasis*; not the naturalized "roadside weed") and alfalfa. The following mixture (per 0.4 hectare [acre] rate) produces outstanding forage quality and also creates favorable bugging sites for grouse and turkeys: 18 kilograms (40 lbs) wheat, 4.5 kilograms (10 lbs) alfalfa, 1.8 kilograms (4 lbs) ladino white clover, and 2.2 kilograms (5 lbs) red clover. Planting recommendations are the same as those described previously. Additional consideration, however, should be given to incorporating lime and fertilizer into what will become the root zone. Ideally, when planting openings, lime (and fertilizers) should be incorporated approximately 15 centimeters (6 in) by plowing/disking for optimum growth (Ball et al. 2002, Harper 2008).

Perhaps the best strategy to provide optimum brood habitat and high-quality forage for grouse and other species in forest openings is to manage for naturally occurring vegetation in the interior of the opening using fire and/or disking, and plant the firebreak around openings in desirable forages. Firebreaks should be approximately two tractor-widths wide and approximately 9–15 meters (30–50 ft) from trees surrounding the opening to allow a soft edge to develop where the forest meets the field. This is especially attractive for grouse using the periphery of an opening, providing easy access to high-quality brood cover, invertebrates, and forages.

Openings should be well distributed throughout management areas to provide/enhance brood habitat and increase interspersion of habitats, which is a critical factor in reducing ruffed grouse home range sizes and possibly leading to increased survival (Fearer 1999, Whitaker 2003, Jones 2005). More, smaller openings will benefit grouse to a much larger extent than fewer larger openings. Realizing other wildlife species are included in most wildlife management plans, larger fields (over 0.8 hectare [2 acres]) can be made more attractive to grouse by breaking the opening into smaller sections using hedgerows comprising soft-mast producers (e.g., crabapple, apple, plum, hawthorn, pear, persimmon, serviceberry, mulberry, dogwood, viburnums, spicebush, elderberry, devil's walkingstick, and Carolina buckthorn). Hedgerows should be relatively wide (9–15 meters [30–50 ft]), not just a single line of trees/shrubs across the field. Instead of merely dividing an opening in two (e.g., two 0.4-hectare [1-acre] sections) with a hedgerow, it may be better to create three or four smaller sections (e.g., four 0.2-hectare [0.5-acre] sections) to create more useable space for grouse with broods.

All forest openings can be made more attractive to grouse by thinning into the forest approximately 100 feet from the edge. As with a wildlife retention cut, percent canopy closure should be reduced to 60–70% (or less) and mast-producing species should be retained, giving emphasis to soft mast producers.

Arrangement and Design of Ruffed Grouse Habitat

Managing habitat for ruffed grouse involves both art and science. Science identifies the reproductive ecology, daily/seasonal movements, home range sizes, survival, and habitat use patterns of grouse. Improving overall habitat quality by arranging cover types in such a way that movements, home range, survival, and reproduction are influenced in a positive manner is an art, guided by a body of scientific knowledge. The creative and skilled manager selects from a variety of methods and techniques to improve habitat quality. Because of variability in forest ownership, management history, management goals, dominant forest types, and other factors, various management approaches must be considered.

Appalachian grouse rely heavily on young forest cover, along with a mix of other cover types, to meet specific seasonal needs. Developing a management design for the array of forest types used by grouse—from young to late rotation and dry uplands to mesic bottoms—may seem impracticable; however, the region's physiography presents unique opportunities to create habitat mosaics preferred by grouse and other wildlife. The topography of the Appalachians creates diverse vegetation communities and associated ecotones (transitional areas between two or more cover types), which often occur in close proximity. Vegetation response differs according to elevation and slope position. Grouse managers should use this intermixing of cover types to their advantage. By planning forest management activities according to topography and associated vegetation, various forest types and age classes are easily interspersed across a management area or landscape. The full benefits of silviculture are realized only when the appropriate methods and techniques are matched with site-specific conditions and habitat objectives. Figure 13.7 shows excellent interspersion of regenerating forest with adjacent thinned and uncut mature forest. Female grouse used this area on the North Carolina study site extensively during fall and winter.

Placement of Habitat Features. Habitats used by ruffed grouse throughout the Appalachians include a variety of forest types and age classes. Young hardwoods 6–20 years old, gated forest roads, mesic stands with an herbaceous understory, and mature mast-producing stands are important habitat features for grouse across the region. To benefit ruffed grouse, a concerted effort must be made to use the appropriate silvicultural techniques and provide the appropriate habitat features in a mosaic across the management area.

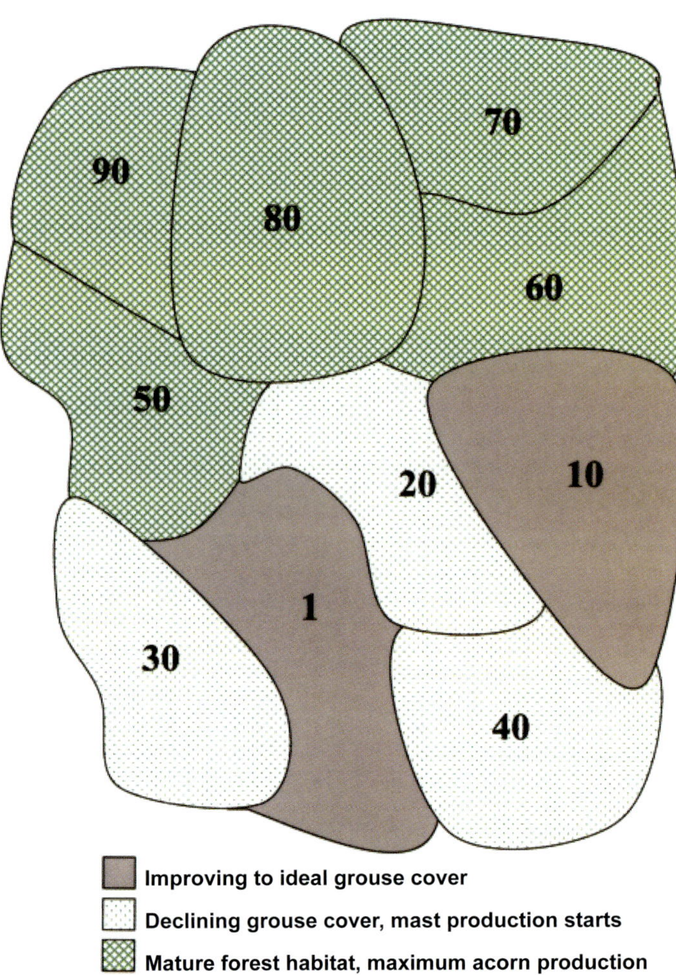

Figure 13.7

Example of a 40-ha (100-acre) oak-hickory woodlot where 10 acres are harvested every 10 years. This creates a wide variety of habitats, from very young forest to old forest, that benefit grouse and many other species of wildlife. The numbers indicate the age of each stand after a full 100-year rotation.

Food should be located adjacent to cover. Forest management planning should ensure mature stands are interspersed with recently regenerated stands to increase food availability, especially acorns, beechnuts, and cherries. For example, a regenerating oak-hickory stand should be adjacent to a mature hard-mast producing (acorns, in particular) stand not scheduled for harvest in the next 40 years. This is especially important if the clearcut method is used because there will be no hard mast available in the harvested unit (as opposed to a shelterwood or shelterwood with reserves harvest that retains some high-quality oak trees). Regenerated mesic stands are best located adjacent to a mature stand of desirable mixed-mesophytic or northern hardwood species, such as yellow or black birch and black cherry, which provide buds and soft mast. Group selection harvests might be positioned within mature mesic stands containing black cherry, serviceberry, and grape with a well-developed understory. Close proximity of brood cover (group selection cuts) with soft mast and herbaceous groundcover would be beneficial for broods through late summer and early fall. In oak-hickory stands, group selection cuts should be placed adjacent to two or three high-quality oaks to help ensure food is available adjacent to cover, while the canopies of those trees are released.

Several high-quality brood sites should be within a relatively small area, or at least connected by corridors of suitable habitat. Hens with broods often use logging roads and riparian zones with a lush herbaceous understory. This linear habitat can link otherwise disjunct habitats. Group selection cuts placed near riparian areas can offer several brood sites within a relatively small area. If positioned appropriately on the landscape, group cuts can provide patches of cover connecting other important habitat and make adjacent mature stands more accessible. Logging roads also can connect brood sites and sources of cover across more xeric stands where cover may be less attractive. Given the average distance traveled during a day by broods, the distance between identified brooding areas should not be farther than 800 meters (about a half-mile; Jones 2005).

Elevation and slope position are important considerations when planning forest management for Appalachian grouse. Higher elevation sites generally have thin soils and are prone to disturbance by wind, ice, and fire. Positioning timber harvests along the mid-slope can increase food and cover resources *and* create corridors for grouse between roosting cover on upper slopes and foraging habitat on lower slopes. When developing prescriptions for mid-slope sites, managers should concentrate on connecting disjunct habitat and providing food, roosting habitat, and cover in close proximity.

Strong consideration should be given to regenerating (or at least thinning) stands on lower slopes, bottomlands, *and along riparian zones,* which is preferred habitat for ruffed grouse during winter and summer if a relatively dense stem density and/or well-developed understory is present. Grouse broods will use bottomland clearcuts and other dense stands (Thompson et al. 1987, Scott et al. 1998, Rusch et al. 2000, Fettinger 2002), as well as mature stands with well-developed understories (Haulton et al. 2003, Jones 2005). Realizing most Best Management Practices are ecologically sound, we believe *excluding forest management from all riparian zones is not sensible* when and where logs can be removed without increasing siltation into the stream. This is especially true along ephemeral and first-order drainages. In some areas, forest management has been entirely excluded far into the uplands because of designated Streamside Management Zone widths, precluding habitat management in areas that may be selected by grouse and other wildlife if the correct structure was present. This policy, where it exists, could be changed to better meet the needs of ruffed grouse, as well as American woodcock and a diversity of other wildlife species.

Logging Road Placement. Placing logging roads adjacent to harvested stands increases interspersion by juxtaposing a food source (forbs and insects on roads) with cover and additional foods (within regenerating stands). However, greatest interspersion is achieved when forest roads go through the middle of harvest units and effectively reduce contiguous stand area, as opposed to traversing only one side. In North Carolina, we documented substantially smaller ruffed grouse home ranges in watersheds where logging roads cut across regenerated stands, as opposed to those watersheds where roads adjoined only one side or end (Jones 2005; Figure 13.3). New logging roads should be planned with this in mind. Likewise, along existing logging roads, stands delineated for harvest should be planned accordingly (above and below the road). Not only does this technique improve grouse habitat, but also skid lanes used to remove logs are shorter because the haul road is in the middle of the stand, not on one end or side. *Increasing interspersion of preferred habitats and positioning cover and food in close proximity across a management area is certainly the most important consideration when managing habitat for ruffed grouse.* Positioning harvest units and logging roads correctly makes this possible.

Size and Shape of Harvest Units. There is a confusing abundance of literature concerning the optimal size of timber harvest units for ruffed grouse (Gullion 1977, Kubisiak et al. 1980, McCaffery et al. 1996, Storm et al. 2003). Recommendations of 0.4–10 hectares (1–25 acres) in mixed oak forests allow good interspersion of young forest patches with other important features. Considering harvesting economics, some managers recommend larger cuts, up to 16 hectares (40 acres) or more. Research has shown grouse will use any size stand (at least some portion of it) large enough to allow regeneration (Sharp 1963, Macdonald et al. 1994, Fearer and Stauffer 2003). However, because interspersion of high-quality habitat within a relatively small area is the most important consideration when managing for ruffed grouse, harvest units should be relatively small (less than 10 hectares [25 acres] with even-aged and two-aged regeneration methods) and well distributed across the management area. This has been demonstrated in Pennsylvania (Storm et al. 2003) where mixed-oak forests managed with 2-hectare (5-acre) patches in a checkerboard arrangement (to maximize interspersion) sustained grouse populations at levels similar to those found in the Lake States.

In the Appalachians, topography and associated moisture gradients strongly influence forest composition. Forest types weave around the mountains in a mosaic according to aspect, elevation, and landform. Of course, operational factors must be considered, but following the natural mosaic of topography when harvesting stands is most sensible in terms of matching regeneration methods with the appropriate forest type. Therefore, harvest units may not be uniform or straight in size or shape, but be adapted naturally on a given site. Following natural patterns in forest structure and composition creates more edge habitat across a management area and helps increase interspersion.

Timber Stand Rotation. Rotation length varies with site, forest type, past intermediate cuttings, and landowner objectives. In terms of financial maturity, rotation length is determined by the growth rate of the stand and local timber markets. When dominant trees cease to respond much to thinning, and height, shoot elongation, and crown expansion have slowed, the stand is normally harvested. For Appalachian hardwoods, this might be as early as 60 years (on better sites) or as long as 120 years on poor sites. Rotation length might be even longer when landowners desire to allow trees to continue to grow very slowly to large diameters. Naturally, this increases the risk of disease and damage by natural factors, and overall timber quality usually declines.

Pressure from special interest groups has caused rotation lengths on national forests to increase to the point that, on a landscape scale, the forest is maturing out of desired age classes for ruffed grouse. Excessive rotation lengths coupled with newly created "zero-cut zones" have decreased land available for improving ruffed grouse habitat significantly on National Forest lands. Where the potential to create valuable young forest cover is limited or eliminated, even greater emphasis must be placed on improving habitat suitability through timber stand improvement practices (e.g., thinning) and, if possible, group selection harvests.

To meet various life requirements and benefit ruffed grouse, management areas in the Appalachians should comprise stands 6–20 years old, well interspersed within mature (>40 years) mast-producing hardwoods or mixed-mesophytic hardwoods, as determined by site. If ruffed grouse were the only consideration, the proportion of various age classes would be determined by that needed to meet the requirements of a grouse population through the year. This is rarely the case. Many other factors, in actuality, dictate the percentage of a forest in various age classes, including aesthetics, timber management considerations, finances, other wildlife species, etc.

High-quality winter foods may be a primary limiting factor in Appalachian grouse populations, especially in landscapes dominated by oak-hickory forests (Whitaker 2003, Devers 2005). Increased forest area of mast-producing age over the past 20 years, however, has **not** led to increased grouse populations in the Appalachians. Providing high quality winter foods *within a stand that also provides high-quality cover* will help ruffed grouse populations increase. This is possible through the shelterwood, irregular shelterwood, and clearcut with reserves methods.

Table 13.1 shows how the distribution of age classes across a forest changes with various rotation lengths. In reality, a number of rotation lengths would be used on a large forest with various forest types and site conditions. Nonetheless, Table 13.1 shows how desirable cover for ruffed grouse represented in the 6–20-year age class is reduced with increasing rotation lengths.

Table 13.1. Percent forest cover within various age classes using three different rotation lengths.

Age Class	60-year Rotation	80-year Rotation	100-year Rotation
1–5	8	6	5
6–20	25	19	15
21–40	33	25	20
>40	33	50	60

In the Appalachians, the mixed-mesophytic and northern hardwood forest types allow rapid growth of yellow-poplar, black cherry, sugar maple, black and yellow birch, American basswood, yellow buckeye, white ash, American beech, and northern red oak. Where additional cover is needed for ruffed grouse, a shorter rotation length (60–70 years) would be prudent for stands in these forest types (Whitaker 2003, Devers 2005). Also, the presence of cherry, birch, and beech reduce the need for acorns in these forests. There was little evidence of nutritional constraint in mesophytic forests that we studied in the Appalachians. Although there is mast potential in mixed-mesophytic and northern hardwood forests, the majority of mast in the Appalachians is produced within the oak-hickory forest type. Longer rotation lengths (80–90 years) within oak-hickory stands allow more time for mast production and may be desirable with regard to variable acorn production across years. All things considered, we believe rotation lengths averaging 80 years would be desirable for ruffed grouse in most areas of the Appalachians.

Perhaps more important than the exact rotation length to providing continued high-quality habitat for ruffed grouse is implementation of intermediate treatments (i.e., thinnings), prescribed burning, and habitat arrangement. A proactive prescription for intermediate thinnings (all forest types) and burning (oak-hickory stands) will enhance the structure and composition of mature and developing stands (20–40 years) for grouse and extend those desirable characteristics of the 6–20-year age class preferred by ruffed grouse in the Appalachians.

In conclusion, ruffed grouse populations can be increased by addressing their habitat needs. Correct habitat management for ruffed grouse in the Appalachians is relatively straightforward: (1) provide adequate early successional forest habitat by incorporating a sensible timber harvest rotation; (2) use regeneration methods that match the site and forest type being regenerated; (3) retain and enhance hard-mast production; (4) manage forest roads and openings in an effective and efficient manner; and (5) address special habitat features, such as soft-mast plantings, seep management, and old homesites, as necessary. Most importantly, however, habitat must be managed in an arrangement that facilitates grouse movements and needs throughout the year. Managing forests to benefit ruffed grouse also will provide habitat used by a variety of other wildlife species that select early-successional habitats. This may be particularly important for a diversity of neotropical migrant songbirds reliant on young forest cover, because many of these species have shown declining populations over the past several decades.

14

Managing Private Forestland

Mark A. Banker

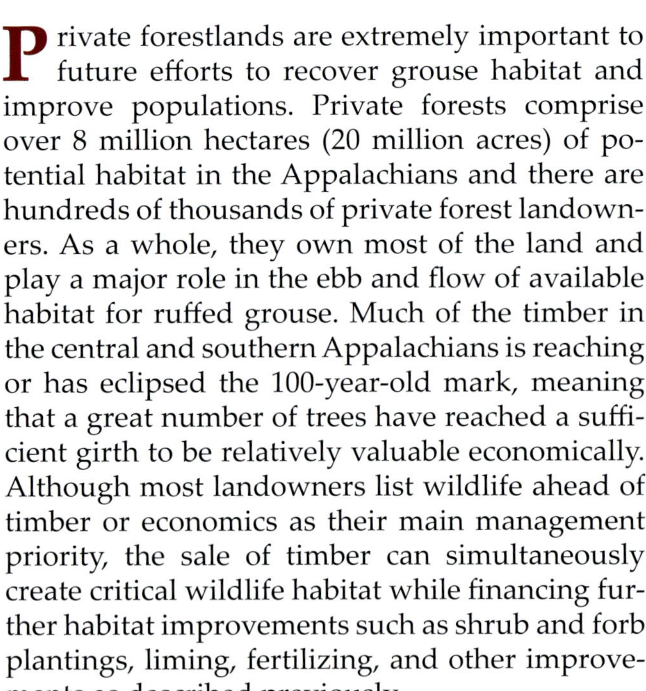

Private forestlands are extremely important to future efforts to recover grouse habitat and improve populations. Private forests comprise over 8 million hectares (20 million acres) of potential habitat in the Appalachians and there are hundreds of thousands of private forest landowners. As a whole, they own most of the land and play a major role in the ebb and flow of available habitat for ruffed grouse. Much of the timber in the central and southern Appalachians is reaching or has eclipsed the 100-year-old mark, meaning that a great number of trees have reached a sufficient girth to be relatively valuable economically. Although most landowners list wildlife ahead of timber or economics as their main management priority, the sale of timber can simultaneously create critical wildlife habitat while financing further habitat improvements such as shrub and forb plantings, liming, fertilizing, and other improvements as described previously.

To make the best use of their land, a landowner should have a game plan. This starts with identifying personal objectives for the land. There are generally two kinds of land management objectives: those that make the landowner personally happy, and those that address a resource need. Personal and resource-oriented goals almost always overlap. For a forest landowner, almost any kind of wildlife-oriented objective will require some form of timber harvest, which will require some careful planning up-front. Before deciding on specific management prescriptions, it is important to get the opinion of a professional forester or biologist, preferably one who has little or no stake in the financial aspects of managing your property, or one who is a licensed consultant. Many times, they can help develop a detailed plan outlining the management of your forest for the foreseeable future and help you implement your goals. Additionally, accessing a Forest Stewardship program can be helpful. Many of the recommendations made in Chapter 13 are applicable to management of private forestlands for ruffed grouse in the Appalachians as well as public lands.

Landowners in the Appalachians are usually going to be managing in oak-dominated forests. If the landowner happens to have forest types other than oak-hickory on their property, many of recommendations for managing oak forests also will apply to other potential forest types. Oak forests can provide important habitats for ruffed grouse and many other types of wildlife. Although grouse generally are more numerous in the aspen-dominated forests of the northern Great Lakes region, they are common in oak forests where suitable young forest habitats are available, as has been shown by the work of the ACGRP.

Dense, young oak forests less than 20 years old provide critical habitat for ruffed grouse, yellow-breasted chats, blue-winged warblers, eastern towhees, and many other songbirds. Grouse and many migratory songbird species are declining throughout the eastern United States as our forests continue to mature, and young forests become uncommon.

Historically, oak forests were maintained by periodic wildfires that raced across the landscape. These fires opened the forest and provided the sunlight required by young oak trees to grow and develop into a new, vigorous oak forest. Today, we no longer allow fire to shape our forests as it has done for millennia. Therefore, to provide the necessary growing conditions, we must proactively manage oak forests through the thoughtful implementation of commercial timber harvest practices. Without such management, oak forests will eventually die out and convert to other types of forest, such as maple, that are less valuable to grouse and other wildlife.

Landowners interested in supporting a diversity of wildlife and maintaining the oak component of their forest should consider the use of wildlife habitat management practices such as a commercial timber harvest. Oak trees can regenerate from seed or from sprouts that arise from the base of the stump of a tree that has been cut or burned. Because young oak trees do not grow well in shade, it is necessary to remove as many of the mature overstory trees as possible within the harvest area. Oak seedlings exposed to full sunlight develop rapidly into a dense, young forest that will protect grouse from hawks, owls, and other predators. Ruffed grouse benefit most from patches of young trees 6–20 years old and 1–16 hectares (2–40 acres) in size.

In hilly terrain, young oak forests growing on north- or east-facing slopes tend to provide the best habitat for grouse. These slopes are not exposed to the heat of the afternoon sun, so they stay relatively cool and moist. Such conditions often support succulent ground vegetation, an important food source for grouse. Positioning a cut low on a slope, especially adjacent to a shrub-dominated old field or near a stream, will provide grouse access to good brood cover and fruit-producing shrubs, such as blackberry, dogwood, viburnum, elderberry, and grape.

Trees can be managed on a rotation that allows the landowner to offer a steady supply of grouse habitat and timber over the long haul. Remember, the most attractive forest habitat for grouse is generally 6–20 years old (young forest). Because the benefits of young oak forests as cover for grouse last only a few years, landowners interested in ruffed grouse should try to maintain oak stands of varying ages, if possible. Oaks in particular can reasonably be managed on a 100-year rotation.

To understand the concept of managing an oak forest on a 100-year rotation, consider a 100-acre tract of forest (Figure 13.7). If you harvest 10% (4 hectares [10 acres]) of that forest every 10 years, you will have 10–20% of your forest in prime grouse habitat for the next 100 years while also offering the full range of habitat values, such as acorns and other hard mast, provided by forests of more advanced age. Of course, the grouse enthusiast not counting on being around for another 100 years can create as much grouse habitat as he wants when he wants.

Besides a timber sale, there are other creative ways for removing overstory trees to allow for a more dense understory attractive to grouse. Fuelwood cutting can be a practical and productive activity. The landowner can cut trees for firewood strategically over time to allow more sunlight into the ground, or mark particular trees that can be cut by friends and family that may need wood. Firewood also can be sold. Girdling a few dominant trees that cast the most shade can be an easy, relatively safe, and effective way to promote undergrowth, as well as to create habitat for wildlife that use snags. The entire circumference of the trees should be cut at least an inch deep to stop the flow of nutrients from the roots to the top of the tree. To insure successful girdling, two complete rings should be cut around the tree. It may take the tree a year or two to succumb entirely to girdling.

The nutritional component of oak forests may be nearly as important as cover. Acorns are a critically important food supply for wild turkeys, white-tailed deer, black bears, squirrels, and the mice and other small mammals that are themselves a primary food source for hawks, owls, fox, and other predators. Ruffed grouse also commonly feed on acorns when available on the forest floor. As previously noted, the ACGRP strongly suggested that acorn availability in the fall is closely tied to grouse reproductive success the following spring. Oak trees that are 40–60 years old and older typically produce more acorns than younger trees. When oaks fail to produce acorns, grouse are forced to endure the winter by consuming much less nutritious foods in oak-dominated forests.

Grouse living in northern hardwood forests (referred to as "mixed-mesophytic" forests elsewhere in this book) at the highest elevations of the southern Appalachians are apparently not as reliant on acorns. A diversity of soft mast and buds provided by species like cherry, birch, maple, and even aspen make acorns less critical.

Harvesting trees is not the only option for improving your private land for grouse. Old fields can also be good grouse cover once they begin to covert to a high percentage of shrubs and saplings. Dense grass fields are generally not used by grouse, but fields that have a mixture of woody cover and small patches of herbaceous vegetation can offer excellent food and cover. Old fields can be maintained by cutting overstory trees that shade shrubs and saplings, regenerating whole patches of trees and shrubs by cutting with chainsaw or heavy mower, or through burning. The process of old field succession from grass to woody cover can be encouraged by planting patches of shrubs such as dogwood, viburnum, hawthorn, or crabapple. Thickets of grape vines can be very attractive to grouse for their food and cover values. Small patches of grape, even single vines, can be expanded by providing more sunlight. Trees in which grapes are growing can be felled into a pile that will eventually become a large grape tangle.

In the Appalachians, large patches of mountain laurel and rhododendron are probably responsible for preventing grouse from disappearing completely from large tracts of mature forest. Laurel and rhododendron thickets offer fair cover, change very slowly, and maintain their leaves throughout the year. These habitats can be particularly important in winter when leaf fall makes other woody habitats less secure. There is no easy way to establish either laurel or rhododendron in appreciable amounts where it is not already present.

Dense, secure cover is usually necessary to consistently keep grouse in an area. Once that cover is established, there are some important ways that landowners can make the habitat even better. The recommendations found earlier in this chapter related to managing habitat on roads and in openings apply well to private lands.

In most cases, landowners should not try to go it entirely alone, particularly with regard to selling timber to make habitat improvements. To avoid costly mistakes, consider getting the assistance of a natural resource professional trained in forest/wildlife management. Done properly, a timber harvest can improve habitat for grouse, protect water quality, allow for future harvests, and establish regeneration while also providing income. Without sound guidance, a single timber harvest can degrade your land and decrease its value for generations. A common practice called "high-grading" or "diameter limit cutting" involves cutting only the largest and most valuable trees. The trees left behind after this practice often have little potential to develop into high-quality trees. Cutting just big trees takes the best ones and leaves the rest. Rather than stripping your forest of its best assets, a good timber harvest leaves your forest in a condition to provide financial and natural benefits down the road.

Many state wildlife and forestry agencies offer special assistance to private forest landowners who want to manage their land specifically for wildlife. This includes developing detailed habitat management plans. Non-profit wildlife organizations, such as the Ruffed Grouse Society (RGS), also offer assistance to landowners free of charge and will work in concert with state resources professionals and consultants to insure a well-rounded approach. In some cases, states will act as third party reviewers when a private consultant develops a management plan, giving the landowner the added security of knowing that the plan is approved by someone with no financial interest in the management.

Maryland, Ohio, Pennsylvania, and West Virginia currently offer intensive training for forest landowners at least once a year through the natural resources extension programs of the respective state universities. These similar programs, which operate under various names, were spawned by the Ruffed Grouse Society's "Coverts" program and are still partially supported by RGS. The training gives the landowner most of the knowledge necessary to successfully manage for grouse. These programs also strongly encourage "graduates" to share their knowledge with other landowners and strive to track the effectiveness of such communication. Throughout the year, workshops and refresher courses are offered in addition to the annual training course. University Natural Resource Cooperative Extension programs usually also have a statewide network of professionals focused on assisting the private landowner.

Forest landowner groups are gaining momentum in some states. Woodland owner groups are independent nonprofit organizations (usually) interested in sound forest management. These groups can give you an opportunity to meet other woodland owners and participate in meetings, educational programs, and property tours. Fellow members also can introduce you to natural resource professionals they have worked successfully with. Pennsylvania, for example, has over 20 such woodland owner groups.

In summary, landowners interested in ruffed grouse, white-tailed deer, and other species that thrive in young forest habitats have the potential to provide these habitats through the use of commercial timber harvest practices or other types of habitat management.

15

The Future of Ruffed Grouse Management in the Real World

Mark A. Banker and Daniel R. Dessecker

Grouse habitat is by nature ephemeral. The dense, young forests that grouse prefer start harboring grouse about six years after disturbance and usually lose their attraction within 20 years. It has been said that the life span of a good grouse covert is about the same as a good bird dog. The longevity of a grouse covert varies by site conditions and tree species. Grouse rely on some sort of disturbance to the forest overstory that allows sunlight to stimulate the growth of woody plants, often tens of thousands of stems per hectare. Before extensive European settlement in the eastern U.S., naturally occurring and Native American-induced fire continually maintained a mosaic of early successional and mature forest habitat on many landscapes in the South (Lorimer 2001). The area of early successional habitat probably fluctuated considerably over time. The agricultural activities of Native Americans likely created additional habitat dominated by small trees or shrubs. Today, most of the forests in the central and southern Appalachians are on their second or third generation following the nearly complete harvest of trees in the late 1800s and early 1900s.

Despite growing concerns over urban sprawl, there is now more forest in the southern Appalachians than there was even a few decades ago. More than 10.6 million hectares (26 million acres) of forest covers 70% of the region (SAMAB Cooperative 1996). Ironically, it is the lack of disturbance in the past 30 years or so that has caused declines in grouse populations. The amount of early successional habitat has declined by 16% since the 1970s, and drumming counts and hunter success rates suggest that grouse numbers have declined dramatically over the same period. A major reason for these unfortunate changes is the almost exclusive reliance on commercial timber harvest to create significant patches of good grouse habitat. Though prescribed burning is gaining some favor for habitat management, it is not practiced on a large enough scale to impact grouse populations. For safety and economic reasons, wildfires are quickly suppressed. For example, from 1988 to 1993, 32,074 wildfires burned 185,382 hectares (454,186 acres) in the central and southern Appalachians, or about 0.3% of the forested acres per year. About 1% of these fires burned more than 40 hectares (100 acres). Additional natural disturbance from uncommon events such as hurricanes and tornadoes offers unpredictable and short-lived benefits.

Our best opportunities to sustain high-quality grouse habitat are on state and federal public lands that are managed by professionals. Careful planning must take place to insure that early successional habitats are maintained or replaced as they naturally grow and change. Such planning often must include input from a broad array of interested publics. The management of mature forests at regular intervals to insure a steady supply of early successional habitat is referred to as a timber rotation. Forests carefully managed on a timber rotation often provide all of the needs of the wildlife that may occur on that particular tract. In the central and southern Appalachians, public land that is regularly managed to provide a steady supply of early successional habitat is uncommon. Recent trends suggest that timber harvesting is on the rise on private lands (USDA Forest Service 2002); there is some indication of modest increases on public lands, as well. However, many of these harvests do not remove sufficient overstory to allow the development of quality early successional habitat.

Much management for grouse boils down to cutting trees in one way or another. The notion of cutting trees elicits a broad range of responses. Many recognize the value of professionally managing forests for an array of wildlife and wood products. Usually, both goals are met at the same time. Others believe that there is no good reason to cut a tree and oppose any form of timber harvest, particularly on public lands.

Our best wildlife science indicates that the thoughtful application of forest management practices (timber harvest) is important for maintaining habitat for a wide diversity of wildlife species (Annand and Thompson 1997, Litvaitis 2001, Fuller and DeStefano 2003). A mosaic of young and mature forests is important for grouse. Similarly, species typically associated with mature forests, especially songbirds, have been observed using early successional habitat in late summer and during migration (Anders et al. 1998, Vega-Rivera et al. 1998b, Pagen et al. 2000, Hunter et al. 2001). Research has shown that forested landscapes managed with a variety of silvicultural techniques to provide a variety of forest age classes support a greater diversity of wildlife than unmanaged forests (Thompson et al. 1992, Duguay et al. 2001, Holmes and Sherry 2001).

Most of the public land and best potential grouse habitat in the Appalachians is managed by the U.S. Forest Service. From western Pennsylvania to northern Georgia, the Forest Service controls about 2.6 million hectares (6.3 million acres) of forestland. Public ownership of all other kinds amounts to about 2.5 million hectares (6 million acres), but less than 0.8 million hectares (2 million acres) south of Pennsylvania. The active management of forests through timber harvest on Forest Service lands has plummeted in the past two decades, mostly in response to repeated legal challenges by special interest groups opposed to such management. This has led to dramatic declines in habitat for grouse and dozens of other game and non-game species. Without the consistent input of hunters and other dedicated conservationists and constant vigilance from forestry and wildlife professionals, forest management on National Forests might have been even more restricted.

Under the National Environmental Policy Act (NEPA), the Forest Service is required to perform an extensive social and ecological analysis of nearly all forms of management. NEPA allows for decisions by the Forest Service to be appealed and then challenged in court. Legal challenges—often based on administrative processes rather than on the ecological merits of forest management—have been used to reverse hundreds of projects that would have created grouse habitat on National Forests over the years. At best, legal challenges, even unsuccessful ones, can tie up management decisions and actions for years. The financial and administrative strain of detailed analysis and/or legal defense of forest management projects, as well as self-imposed limitations on timber harvest, have led to the dramatic decline in early successional habitat on National Forests. State-owned forestland tends to be smaller and more intensively managed than federal lands. State lands are largely not encumbered by the same degree of regulation. Regardless of ownership, other factors such as timber markets, access, human resources, weather, and budgets can influence the progress of forest management.

Closely tied to the reduction in forest management on public lands is oak decline. Though oak forests dominate the rural landscapes of the Appalachians, the process of oak decline is slowly changing forest composition. Oak-hickory and oak-pine forests cover more than 6.9 million hectares (17 million acres) of the central and southern Appalachians, making it about six times more abundant than the next most-common forest type, southern yellow pine (SAMAB 1996). Forest inventory data document that oak forests in the Appalachians are, when taken as a whole, growing older as existing stands continue to mature and few new, young stands become established.

The ACGRP (Norman et al. 2004) strongly suggested that successful reproduction and recruitment of young grouse into the fall population is closely tied to acorn crops the previous year in oak-dominated forests. Turkey, deer, bear, and a diversity of other species also depend on acorns. Acorns and other seeds represent the most valuable and energy-rich plant food available to eastern wildlife in the dormant season. Acorn production can be 3–10 times greater than browse production in oak forests. Oak has increased in importance as a food for wildlife because of the elimination of the American chestnut and decline in American beech. A focus on oak management is important not only because of its value to wildlife, but also because of its dependence on disturbance for regeneration (Thompson and Dessecker 1997).

The natural process of oak decline is a result of multiple factors, including environmental stress, root disease, and insect pests of opportunity. The result is a slow, progressive dieback of overstory trees from the top down and from the outside inward. Oak decline is more prominent on national forestlands than on other ownerships due to the higher frequency of oak-dominated stands of advanced physiological age on sites with average to low site productivity. Without management that removes the mature overstory and allows oak seedlings to grow rapidly in sunlight, oaks eventually are replaced by more shade-tolerant tree species (such as red maple and black gum) that are of less value to wildlife. The decline in forest management has led to extensive tracts of very mature oak with very little regenerating, younger oak forests. Essentially, there is no mechanism for initiating the next generation of oak forests (SAMAB 1996).

Maintenance of high-quality grouse brood habitat in forested landscapes is another critical management challenge. One of the more ominous discoveries of the ACGRP was the poor survival of grouse chicks in the region. Fewer than 20% of grouse chicks survived beyond five weeks. Grouse chicks feed nearly exclusively on insects for the first month after hatching. In the extensive forests of the Appalachians, the herbaceous vegetation that supports insects is a crucial, but often scarce, habitat component. Brood habitat is characterized by a combination of dense woody cover with an herbaceous component nearby. Small, well-distributed herbaceous openings and seeded, lightly or un-traveled roads make high-quality brood habitat. Timber stand improvements that remove enough overstory to allow sunlight to reach the forest floor and encourage lush herbaceous growth will also be used by grouse broods. Forest management followed by the seeding of log landings and roads is an efficient way to create brood habitat. The decline in all forms of forest disturbance has likely affected the availability of brood habitat. Declining wildlife budgets also have led some land managers to abandon the maintenance of wildlife openings and roads in some remote forested tracts.

Riparian, or streamside zones, offer unique habitat benefits for grouse and other wildlife, but are seldom managed with terrestrial wildlife in mind under modern guidelines. The normally steep, mountainous topography of the Appalachians gives rise to a complex arrangement of small and medium-sized perennial, intermittent and ephemeral streams that offer important wildlife benefits and management challenges. Streamside zones, even adjacent to ephemeral streams that may only carry water for a month or so each year, are often off limits to forest management to varying degrees. The ACGRP showed that mesic bottomland (riparian) habitats are important foraging areas for hens in oak-hickory forests, probably due to the relative abundance of herbaceous plant foods promoted by relatively moist and fertile soils. Greater availability of mesic bottomlands was associated with smaller home ranges of hens, and smaller home ranges were positively associated with successful reproduction. The availability of herbaceous plants in riparian zones on the forest floor is dependent on sunlight. Even the removal of scattered, big-canopied trees in a streamside zone can greatly enhance habitat quality for grouse in vast oak-hickory forests.

Uncertainty and untested assumptions about the effects of forest management on water quality has led to the nearly complete avoidance of streamside zones by land managers. The effects of forest management in streamside zones has not been extensively studied; nonetheless, research at locations such as the Coweeta Hydrologic Laboratory in southwestern North Carolina has suggested that many common assumptions about negative impacts on water quality downstream from intensive forest management may not hold up under scientific scrutiny (Swank et al. 2001). Although riparian areas unquestionably warrant special protections, broad-brush policies that preclude habitat management activities from these sites are ecologically unjustified and potentially exacerbate declines of ruffed grouse, American woodcock, and other wildlife of young forests.

Fortunately, some grouse habitat in the central and southern Appalachians does not require any kind of management. Naturally occurring mountain laurel and rhododendron thickets can offer fair grouse habitat in otherwise mature forests, especially in winter. The evergreen and slowly changing nature of mountain laurel and rhododendron can provide a stable, if not optimal, alternative to regenerating young forests for many decades. Some grouse have been observed occupying young, regenerating forest in the summer and then traveling some distance to spend the winter in mountain laurel thickets.

Public support for professional management of public forestland is crucial for the future of grouse management and grouse hunting in the Appalachian Mountains. Research conducted by the ACGRP and other scientists offers strong support for future efforts to stabilize and increase the availability of habitat for grouse and many other species of wildlife. Measures to simplify the process of managing federal lands need to be implemented. Support from hunters and other conservationists will continue to be crucial.

The vast acreage of forestland in the southern Appalachians suggests that the downward trend in ruffed grouse habitat and populations is reversible. Habitat for grouse on private lands is likely to increase as trees continue to mature and increase in value and are harvested. We must continue to work with private forest landowners wherever possible. So far, private lands management has not balanced out the lack of management elsewhere, nor are most private lands professionally managed or open to the public.

Status quo on National Forests will result in continued declines in grouse on public lands. A more balanced approach to management will have to evolve rapidly in order to maintain healthy grouse populations in places where the average person can enjoy seeing, hearing, and hunting them. Increased prescribed burning can improve understory characteristics beneficial to grouse and compliment early successional habitat. The commercial timber harvest program must be reinvigorated for the sake of many wildlife species. A major shift in philosophy by traditional environmental groups that supports wildlife management on public lands will be very helpful for implementing needed changes in the forested landscape.

The Ruffed Grouse Society

In 1961, three business associates from Monterey, Virginia, noticed declining numbers of ruffed grouse in their Virginia forests and decided to start an organization to draw attention to, and provide assistance for, their treasured upland game bird. With a little effort, they were able to enlist the help of other grouse enthusiasts and soon the Ruffed Grouse Society (Society) was established.

The Society is an international wildlife conservation organization headquartered in Coraopolis, Pennsylvania (Pittsburgh). It employs a team of wildlife biologists to work with local, state, and federal government landowners, as well as private landowners who are interested in improving their land for ruffed grouse, woodcock, and other wildlife that have similar habitat requirements. The Society provides individualized attention to landowners, works with agencies to expand habitat on public lands, and supports applied research projects. The Society has implemented hundreds of habitat projects that affect hundreds of thousands of acres throughout the range of the ruffed grouse and American woodcock.

The Society also has a team of regional directors who help organize chapters consisting of members who share a common interest in these birds and other wildlife. These chapters help support land management projects that improve habitat and hunting conditions. The chapters also sponsor banquets that bring people with common interests together socially to honor their peers and to raise funds to support our conservation programs.

—Mark A. Banker

To learn more about the Ruffed Grouse Society, visit:

www.ruffedgrousesociety.org

Theses and Dissertations
Resulting from the Appalachian Cooperative Grouse Research Project

Bumann, G.B. 2002. Factors influencing predation on ruffed grouse in the Appalachians. MS Thesis, Virginia Tech, Blacksburg, VA, USA.

Devers, P.K. 2005. Population ecology of and the effects of hunting on ruffed grouse *(Bonasa umbellus)* in the southern and central Appalachians. PhD Dissertation, Virginia Tech, Blacksburg, VA, USA.

Dobony, C.A. 2000. Factors influencing ruffed grouse productivity and chick survival in West Virginia. MS Thesis, West Virginia University, Morgantown, WV, USA.

Endrulat, E.G. 2003. The effects of forest management on home range size and habitat selection of ruffed grouse in Rhode Island, Virginia, and West Virginia, USA. MS Thesis, University of Rhode Island, Kingston, RI, USA.

Fearer, Todd M. 1999. Relationship of ruffed grouse home range size and movement to landscape characteristics in Southwestern Virginia. MS Thesis, Virginia Tech, Blacksburg, VA, USA.

Fettinger, J.L. 2002. Ruffed grouse nesting ecology and brood habitat in western North Carolina. MS Thesis, University of Tennessee, Knoxville, TN, USA.

Haulton, G.S. 1999. Ruffed grouse natality, chick survival, and brood micro-habitat selection in the southern Appalachians. MS Thesis, Virginia Tech, Blacksburg, VA, USA.

Jones, B.C. 2005. Ruffed grouse habitat use, reproductive ecology, and survival in western North Carolina. PhD Dissertation, University of Tennessee. Knoxville, TN, USA.

Long, C.R. 2007. Pre-breeding food habitats and condition of ruffed grouse and effects on reproduction in the central and southern Appalachians. MS Thesis, West Virginia University, Morgantown, WV, USA.

O'Keefe, J.M. 2004. Use of elevation by ruffed grouse in Virginia and Pennsylvania. MS Thesis, Eastern Kentucky University, Richmond, KY, USA.

Plaugher, G.F. 1998. Seasonal habitats, foods, and movements of ruffed grouse in the central Appalachian Mountains of West Virginia. MS Thesis, West Virginia University, Morgantown, WV, USA.

Proctor, A. B. 2010. Effect of nutritional deficiency on ruffed grouse condition and reproductive success. MS Thesis, West Virginia University, Morgantown, WV, USA.

Schumacher, C.L. 2002. Ruffed grouse habitat use in western North Carolina. MS Thesis, University of Tennessee, Knoxville, TN, USA.

Smith, B.W. 2006. Nesting Ecology, Chick Survival, and Juvenile Dispersal of ruffed grouse *(Bonasa umbellus)* in the Appalachian Mountains. PhD Dissertation, West Virginia University, Morgantown, WV, USA.

Tirpak, J.M. 2000. Influence of microhabitat structure on nest success and brood survival of ruffed grouse in the central and southern Appalachians. MS Thesis, California University of Pennsylvania, California, Pennsylvania, USA.

Tirpak, J.M. 2005. Modeling ruffed grouse populations in the central and southern Appalachians. PhD Dissertation, Fordham University, Bronx, NY, USA.

Whitaker, D.M. 2003. Ruffed grouse *(Bonasa umbellus)* habitat ecology in the central and southern Appalachians. PhD Dissertation, Virginia Tech., Blacksburg, Virginia, USA.

Common and Scientific Names of Animals and Plants Referred to in the Text

Mammals

Bat, Big Brown	*Eptesicus fuscus*
Bat, Hoary	*Lasiurus cinereus*
Bat, Red	*Lasiurus borealis*
Bison	*Bison bison*
Black Bear	*Ursus amreicana*
Bobcat	*Lynx rufus*
Chipmunk	*Tamias striatus*
Coyote	*Canis latrans*
Deer mouse	*Peromyscus maniculatus*
Dog	*Canis familiaris*
Eastern Chipmunk	*Tamias striatus*
Fisher	*Martes pennanti*
Fox, Gray	*Urocyon cinereoargenteus*
Fox, Red	*Vulpes vulpes*
Groundhog	*Marmota monax*
House Cat	*Felis catus*
Meadow Vole	*Microtus pennsylvanicus*
Mink	*Mustela vison*
Muskrat	*Ondatra zibethicus*
Opossum	*Dildelphis virginiana*
Rabbit	*Sylvilagus floridanus*
Raccoon	*Procyon lotor*
Short-tailed shrew	*Blarina brevicauda*
Striped Skunk	*Mephitis mephitis*
Weasel	*Mustela* spp.
Weasel, Long-tailed	*Mustela frenata*
White-footed Mouse	*Peromyscus leucopus*
White-tailed Deer	*Odocoileus virginianus*

Birds

American Woodcock	*Scolopax minor*
Black Duck	*Anas rubripes*
Blue Jay	*Cyanocitta cristata*
Brown Thrasher	*Toxostoma rufum*
Common Yellowthroat	*Geothlypis trichas*
Dark-eyed Junco	*Junco hyemalis*
Dickcissel	*Spiza americana*
Eagle, Bald	*Haliaeetus leucocephalus*
Eagle, Golden	*Aquila chrysaetos*
Eastern Meadowlark	*Sturnella magna*
Eastern Towhee	*Pipilo erythrophthalmus*
Gray Catbird	*Dumetella carolinensis*
Grouse, Black	*Tetrao tetrix*
Grouse, Blue	*Dendragapus obscurus*
Grouse, Red	*Lagopus lagopus scotius*
Grouse, Ruffed	*Bonasa umbellus*
Grouse, Sage	*Centrocercus urophasianus*
Hawk, Broad-winged	*Buteo platyperus*
Hawk, Cooper's	*Accipiter cooperii*
Hawk, Red-tailed	*Buteo jamaicensis*
Hawk, Red-shouldered	*Buteo lineatus*
Heath Hen	*Tympanuchus cupido cupido*
Indigo Bunting	*Passerina cyanea*
Northern Bobwhite	*Colinus virginianus*
Quail	*Colinus virginianus*
Northern Goshawk	*Accipiter gentilis*
Owl, Barred	*Strix varia*
Owl, Great Horned	*Bubo virginianus*
Owl, Saw-whet	*Aegolius acadicus*
Ptarmigan, Rock	*Lagopus muta*
Ptarmigan, White-tailed	*Lagopus lecura*
Ptarmigan, Willow	*Lagopus lagopus*
Sparrow, Grasshopper	*Ammodramus savannarum*
Sparrow, Henslow's	*Ammodramus henslowii*
Warbler, Blue-winged	*Vermivora pinus*
Warbler, Chestnut-sided	*Dendroica pensylvanica*
Wild Turkey	*Meleagris gallopavo*
Yellow-breasted Chat	*Icteria virens*

Reptiles and Amphibians

Black Rat Snake	*Elaphe obsoleta*

Invertebrates

Ants, Wasps, and Bees	Order Hymenoptera
Beetles	Order Coleoptera
Cicadas, Leafhoppers, Aphids	Order Homoptera
Aphids	Order Hemiptera
Flies	Order Diptera
Spiders and Ticks	Order Arachnida
True Bugs	Order Hemiptera
Gypsy Moth	*Lymantria dispar*

Trees, Shrubs and Woody Vines

Apple	*Malus pumila*
Ash, White	*Fraxinus americana*
Aspen	*Populus* spp.
Aspen, Trembling	*Populus tremuloides*
Aspen, Bigtooth	*Populus grandidentata*
American Beech	*Fagus grandifolia*
Azalea	*Rhododendron* spp.
Basswood	*Tilia americana*
Birch	*Betula* spp.
Birch, Black	*Betula lenta*
Birch, Sweet	*Betula lenta*
Birch, Yellow	*Betula alleghaniensis*
Blackberry	*Rubus* spp.
Black Locust	*Robinia pseudoacacia*
Black Gum	*Nyssa sylvatica*
Blueberry	*Vaccinium* spp.
Buffalo Nut	*Pyrularia pubera*
Carolina Buckthorn	*Rhamnus caroliniana*
Cherry, Black	*Prunus serotina*
Cherry, Pin	*Prunus pensylvanica*
Crabapple	*Malus* spp.
Cucumbertree	*Magnolia acuminata*
Devil's Walkingstick	*Aralia spinosa*
Dogwood	*Cornus* spp.
Dogwood, Flowering	*Cornus florida*
Eastern Red Cedar	*Juniperus virginiana*
Elderberry	*Sambuchus* spp.
Flame Azalea	*Rhododendron calendulaceum*
Grape	*Vitus* spp.
Greenbrier	*Smilax* spp.
Hawthorn	*Crataegus* spp.
Hemlock, Eastern	*Tsuga canadensis*
Hickory	*Carya* spp.
Hickory, Bitternut	*Carya cordiformis*
Hickory, Mockernut	*Carya tomentosa*
Hickory, Pignut	*Carya glabra*
Hickory, Shagbark	*Carya ovata*
Hornbeam	*Ostrya virginiana*
Huckleberry	*Gaylussacia* spp.
Maple, Red	*Acer rubrum*
Maple, Striped	*Acer pensylvanicum*
Maple, Sugar	*Acer saccharum*
Mountain Holly	*Ilex montana*
Mountain Laurel	*Kalmia latifolia*
Mulberry	*Morus* spp.
Oak	*Quercus* spp.
Oak, Black	*Quercus velutina*
Oak, Blackjack	*Quercus marilandica*
Oak, Burr	*Quercus macrocarpa*
Oak, Chestnut	*Quercus prinus*
Oak, Chinkapin	*Quercusl muehlenbergii*
Oak, Northern Red	*Quercus rubra*
Oak, Pin	*Quercus palustris*
Oak, Post	*Quercus stellata*
Oak, Scarlet	*Quercus coccinea*
Oak, Southern Red	*Quercus falcata*
Oak, Swamp White	*Quercus bicolor*
Oak, White	*Quercus alba*
Pear	*Pyrus* spp.
Pine	*Pinus* spp.
Pine, Pitch	*Pinus rigida*
Pine, Virginia	*Pinus virginiana*
Pine, Table Mountain	*Pinus pungens*
Pine, White	*Pinus strobus*
Persimmon	*Diospyros virginiana*
Plum	*Prunus* spp.
Raspberry	*Rubus* spp.
Rhododendron, Great	*Rhododendron maximum*
Rose	*Rosa* spp.
Rose, Multiflora	*Rosa multiflora*
Serviceberry	*Amelanchier* spp.
Sourwood	*Oxydendrum arboreum*
Spicebush	*Lindera benzoin*
Sumac	*Rhus* spp.
Virburnum	*Viburnum* spp.
Witch-hazel	*Hamamelis virginiana*
Yellow Buckeye	*Aesculus octandra*
Yellow Poplar	*Liriodendron tulipifera*

Herbaceous Plants

Alfalfa	*Medicago sativa*
Avens	*Geum* spp.
Beaked Panicum	*Panicum anceps*
Birdsfoot Trefoil	*Lotus corniculatus*
Black-eyed Susan	*Rudbeckia hirta*
Bluegrass	*Poa* spp.
Bluestem, Big	*Andropogon gerardi*
Bluestem, Broomsedge	*Andropogon virginicus*
Bluestem, Little	*Andropogon scoparius*
Bluet	*Houstonia caerulea*
Brome Grass	*Bromus* spp.
Butterfly Milkweed	*Asclepias syriaca*
Chicory	*Chicorium intybus*
Cinquefoil	*Potentilla* spp.
Clover	*Trifolium* spp.
Coltsfoot	*Tussilago farfara*
Deertongue	*Dichanthelium clandestinum*
Dewberry	*Rubus hispidus*
Evening Primrose	*Oenothera* spp.
Fern, Christmas	*Polystichum acrostichoides*
Fern, Wood	*Dryopteris camyloptera*
Firepink	*Silene virgniica*
Fleabane	*Erigeron* spp.
Goldenrod	*Solidago* spp.
Hawkweed	*Hieracium* spp.
Indiangrass	*Sorghastrum nutans*
Japangrass	*Microstegium vimineum*
Lanceleaved Coreopsis	*Coreopsis lanceolata*
Lyre-leaved Sage	*Salvia lyrata*
Mint	*Mentha* spp.
New England Aster	*Symphyotrichum novae-angliae*
Oats	*Avena* spp.
Orchard grass	*Dactylis glomerata*
Partridge Pea	*Chamaecrista fasciculata*
Partridgeberry	*Mitchella repens*
Pokeberry	*Phytolacca americana*
Poverty Oatgrass	*Danthonia spicata*
Purple Coneflower	*Echinacea* spp.
Purpletop	*Tridens flavus*
Ragwort	*Senecio* spp.
Robin's Plantain	*Erigeron pulchellus*
Roundhead Lespedeza	*Lespedeza capitata*
Rye	*Elymus* spp.
Sideoats Grama	*Bouteloua curtipendula*
Sorrel	*Rumex acetosella*
Strawberry	*Fragaria* spp.
Switchgrass	*Panicum virgatum*
Tall Fescue	*Festuca arundinacea*
Timothy	*Phleum pratense*
Trailing Arbutus	*Epigaea repens*
Wheat	*Triticum aestivum*
Wild Bergamot	*Monarda fistulosa*
Wild Strawberry	*Fragaria* spp.
Wildrye, Canada	*Elymus canadensis*
Wildrye, Virginia	*Elymus virginicus*
Wintergreen	*Gaultheria procumbens*

References

Aebischer, N.J., P.A. Robertson, and R.E. Kenward. 1993. Compositional analysis of habitat use from animal radio tracking data. Ecology 74:1313–1325.

Anders, A.D., J.A. Faaborg, and F.R. Thompson III. 1998. Postfledgling dispersal, habitat use and home range size of juvenile wood thrushes. *Auk* 115:349–358.

Anderson, D.R. and K.P. Burnham. 1976. Population ecology of the mallard VI: the effects exploitation on survival. U.S. Fish and Wildlife Service Resource Publication, Washington DC, USA.

Annand, E.M. and F.R. Thompson III. 1997. Forest bird response to regeneration practices in central hardwood forests. *Journal of Wildlife Management.* 61:159–171.

Archibald, H.L. 1975. Temporal patterns of spring space use by ruffed grouse. *Journal of Wildlife Management* 39:472–481.

———. 1976. Spatial relationships of neighboring male ruffed grouse in spring. *Journal of Wildlife Management* 40:750–760.

Baines, D. and H. Linden. 1991. The impact of hunting on grouse population dynamics. Ornis Scandivacia 23:245–246.

Ball, D.M., C.S. Hoveland, and G.D. Lacefield. 2002. Southern forages. Potash and Phosphate Institute and the Foundation for Agronomic Research. Norcross, Georgia, USA.

Barber, H.L., F.J. Brenner, R. Kirkpatrick, F.A. Servello, D.F. Stauffer, and F.R. Thompson. 1989. Food. Pages 268–282 *in* Atwater, S., and J. Schnell, editors. *The wildlife series: ruffed grouse.* Stackpole Books, Harrisburg, Pennsylvania, USA.

Basinger, R.G. 2003. Silvicultural prescriptions to enhance habitat for wild turkeys in mixed hardwood forests. M.S. Thesis. University of Tennessee. Knoxville, TN, USA.

Beck, D.E. and R.M. Hooper. 1986. Development of a southern Appalachian hardwood stand after clearcutting. *Southern Journal of Applied Forestry* 10:168–172.

Beckerton, P.R. and A.L.A. Middleton. 1982. Effects of dietary protein levels on ruffed grouse reproduction. *Journal of Wildlife Management* 46:569–579.

Begin, J.J. and W.M. Insko, Jr. 1972. The effects of dietary protein level on the reproductive performance of Coturnix breeder hens. *Poulty Science* 51:1662–1669.

Bent, A.C. 1937. Life histories of North American birds of prey. U.S. National Museum Bulletin 167. Washington DC, USA.

Bergerud, A.T. 1985. *Grouse of the World.* University of Lincoln Press, Lincoln, Nebraska, USA.

———. 1988. Population ecology of North American grouse. Pages 578–685 *in* A.T. Bergerud and M.W. Gratson, editors. *Adaptive strategies and population ecology of northern grouse Volume II: Theory and synthesis.* University of Minnesota Press, Minneapolis, USA.

——— and W.M. Gratson, editors. 1988. *Adaptive strategies and population ecology of northern grouse.* University of Minnesota Press, Minneapolis, Minnesota, USA.

Berner, A. and L.W. Gysel. 1969. Habitat analysis and management considerations for ruffed grouse for a multiple use area in Michigan. *Journal of Wildlife Management* 33:769–778.

Boag, D.A. 1976. Influence of changing grouse density and forest attributes on the occupancy of a series of potential territories by male ruffed grouse. *Canadian Journal of Zoology* 54:1727–1736.

Braun, E.L. 1950. *Deciduous forests of eastern North America.* Blakiston Company, Philadelphia, PA.

Brose, P.H. and D.H. Van Lear. 1998. Responses of hardwood advance regeneration to seasonal prescribed fires in oak-dominated shelterwood stands. *Canadian Journal of Forest Research* 28:331–339.

———, ———, and R. Cooper. 1999a. Using shelterwood harvests and prescribed fire to regenerate oak stands on productive upland sites. *Forest Ecology and Management* 113:125–141.

———, ———, and P.D. Keyser. 1999b. A shelterwood-burn technique for regenerating oak stands on productive upland sites in the Piedmont region. *Southern Journal of Applied Forestry.*

Broseth, H. and H.C. Pedersen. 2000. Hunting effort and game vulnerability studies on a small scale: a new technique combining radio-telemetry, GPS, and GIS. *Journal of Applied Ecology* 37:182–190.

Bumann, G.B. 2002. Factors influencing predation on ruffed grouse in the Appalachians. MS Thesis, Virginia Tech, Blacksburg, VA, USA.

——— and D.F. Stauffer. 2004. Predation on adult ruffed grouse in the Appalachians. Pages 35–38 *in* G.W. Norman, D.F. Stauffer, J. Sole, T.J. Allen, W.K. Igo, S. Bittner, J. Edwards, R.L. Kirkpatrick, W.M. Giuliano, B. Tefft, C. Harper, D. Buehler, D. Figert, M. Seamster, and D. Swanson, editors, Ruffed grouse ecology and management in the Appalachian region: Final project report of the Appalachian Cooperative Grouse Research Project. Richmond, VA, USA.

Bump, G., R.W. Darrow, F.C. Edminster, and W.F. Crissey. 1947. *The ruffed grouse: life history, propagation, management.* Holling Press, Buffalo, New York, USA.

Clark, M.E. 1996. Movements, habitat use, and survival of ruffed grouse *(Bonasa umbellus)* in Northern Michigan. MS Thesis, Michigan State University, East Lansing, MI, USA.

———. 2000. Survival, fall movements, and habitat use of hunted and non-hunted ruffed grouse in northern Michigan. PhD Dissertation, Michigan State University. A East Lansing, MI, USA.

Clarke, A.L., B.E. Sæther, and E. Røskaft. 1997. Sex biases in avian dispersal: a reappraisal. *Oikos* 79:429–438.

Clobert, J., J.O. Wolff, J.D. Nichols, E. Danchin, and A.A. Dhondt. 2001. Introduction, p. xvii–xxi. *In* J. Clobert, E. Danchin, A.A. Dhondt, and J.D. Nichols (eds.), *Dispersal*. Oxford University Press, NY.

Clugston, D.A., J.R. Longcore, D.G. McAuley, and P. Dupuis 1994. Habitat use and movements of postfledging American black ducks *(Anas rubiripes)* in the St. Lawrence estuary, Quebec. *Canadian Journal of Zoology* 72:2100–2104.

Conradt, L., P.A. Zollner, T.J. Roper, K. Frank, and C.D. Thomas. 2003. Foray search: an effective systematic dispersal strategy in fragmented landscapes. *The American Naturalist* 161:905–915.

Dale, M.E., H.C. Smith, and J.N. Pearcy. 1995. Size of clearcut opening affects species composition, growth rate, and stand characteristics. USDA Forest Service, Research Paper NE-698, Northeast Forest Experiment Station, Radnor, Pennsylvania, USA.

Darrow, R. 1939. Seasonal food preferences of adult and young grouse in New York State. Transactions of the North American Wildlife Conference 4:585–590.

Davis, J.A. 1969. Aging and sexing criteria for Ohio ruffed grouse. *Journal of Wildlife Management* 33:628–636.

——— and R.J. Stoll. 1973. Ruffed grouse age and sex ratios in Ohio. *Journal of Wildlife Management* 37:133–141.

Dessecker, D.R. and D.G. McAuley. 2001. Importance of early successional habitat to ruffed grouse and American woodcock. Wildlife Society Bulletin 29:456–465.

Destefano, S. and D.H. Rusch. 1986. Harvest rates of ruffed grouse in northeastern Wisconsin. *Journal of Wildlife Management* 50:361–367.

Devers, P.K. 2005. Population ecology of and the effects of hunting on ruffed grouse *(Bonasa umbellus)* in the southern and central Appalachians. PhD Dissertation, Virginia Tech. Blacksburg, VA, USA.

Dickerson, J., B. Wark, D. Burgdorf, T. Bush, R. Maher, C. Miller, B. Poole. No date. Vegetating with native grasses in northeastern North America. Natural Resources Conservation Service Plant Materials Program and Ducks Unlimited Canada.

Dimmick, R.W., M.J. Gudlin, and D.F. McKenzie. 2002. The northern bobwhite conservation initiative. Miscellaneous publication of the Southeastern Association of Fish and Wildlife Agencies, South Carolina.

Dobony, C.A. 2000. Factors influencing ruffed grouse productivity and chick survival in West Virginia. MS Thesis, West Virginia University, Morgantown, WV, USA.

———, J.W. Edwards, W.M. Ford, and T.J. Allen. 2003. Nesting success of ruffed grouse in West Virginia. Proceedings of the Annual Conference of the Southeastern Association of Fish and Wildlife Agencies 55:456–465.

Doerr, P.D., L.B. Keith, D.H. Rusch, and C.A. Fischer. 1974. Characteristics of winter feeding aggregations of ruffed grouse in Alberta. *Journal of Wildlife Management* 38:601–615.

Dorney, R.S. 1963. Sex and age structure of Wisconsin ruffed grouse populations. *Journal of Wildlife Management* 27:599–603.

——— and C. Kabat. 1960. Relation of weather, parasitic disease and hunting to Wisconsin ruffed grouse populations. Wisconsin Conservation Department Technical Bulletin 20, Madison, Wisconsin, USA.

Donahue, R.L, R.W. Miller, and J.C. Shickluna. 1983. *Soils: An introduction to soils and plant growth.* Prentice Hall, Inc. Englewood Cliffs, New Jersey, USA.

Duguay, J.P., P.B. Wood, and J.V. Nichols. 2001. Songbird

abundance and avian nest survival rates in forests fragmented by different silvicultural treatments. *Conservation Biology* 15:1405–1415.

Dykes, S.A. 2005. Effectiveness of native grassland restoration in restoring grassland bird communities in Tennessee. MS Thesis, University of Tennessee. Knoxville, TN, USA.

Elliott, K.J., S.L. Hitchcock, and L. Krueger. 2002. Vegetation response to a large scale disturbance in a southern Appalachian forest: Hurricane Opal and salvage logging. *Journal of the Torrey Botanical Society* 129:48–59.

——— and W.T. Swank. 1994. Changes in tree species diversity after successive clearcuts in the southern Appalachians. *Vegetation* 115:11–18.

Endrulat, E.G. 2003. The effects of forest management on ruffed grouse home range size and habitat selection in Rhode Island, Virginia, and West Virginia, USA. MS Thesis, University of Rhode Island, Kingston, RI, USA.

Epperson, RG. 1988. Population status, movements and habitat utilization of ruffed grouse on the Catoosa Wildlife Management Area, Cumberland County, Tennessee. MS Thesis, University of Tennessee Knoxville, Knoxville, TN, USA.

Errington, P.L. and F.N. Hamerstrom, Jr. 1935. Bobwhite winter survival on experimentally shot and unshot areas. Iowa State College Journal of Science 9:625–639.

Fearer, T.M. 1999. Relationship of ruffed grouse home range size and movement to landscape characteristics in southwestern Virginia. MS Thesis, Virginia Tech, Blacksburg, Virginia, USA.

——— and D.F. Stauffer. 2003. Relationship of ruffed grouse *(Bonasa umbellus)* home range size to landscape characteristics. *American Midland Naturalist* 150:104–114.

——— and ———. 2004. Relationship of ruffed grouse *Bonasa umbellus* to landscape characteristics in southwest Virginia, USA. *Wildlife Biology* 10:81–89.

Fettinger, J.L. 2002. Ruffed grouse nesting ecology and brood habitat in western North Carolina. Thesis, University of Tennessee. Knoxville, Tennessee. 105 pp.

———, C.A. Harper, and C.E. Dixon. 2002. Invertebrate availability for upland game birds in tall fescue and native warm-season grass fields. *Journal of the Tennessee Academy of Science* 77:83–87.

Fischer, C.A. and L.B. Keith. 1974. Population responses of central Alberta ruffed grouse to hunting. *Journal of Wildlife Management* 38:585–600.

Fischer, R.A., A.D. Apa, W.L. Wakkinen, and K.P. Reese. 1993. Nesting-area fidelity of sage grouse in southeastern Idaho. *Condor* 95:1038–1041.

Frost, C.C. 1998. Pre-settlement fire frequency regimes of the United States: A first approximation. Tall Timbers Fire Ecology Proceedings 20:70–81.

Fuller, T.K. and S. DeStefano. 2003. Relative importance of early-successional forests and shrubland habitats to mammals in the northeastern United States. *Forest Ecology and Management.* 185:75–79.

Gandon, S., and Y. Michalakis. 2001. Multiple causes of the evolution of dispersal, p. 155–167. *In* J. Clobert, E. Danchin, A.A. Dhondt, and J.D. Nichols (eds.), *Dispersal*. Oxford University Press, NY.

Gardner, F.A. and L.L. Young. 1972. The influence of dietary protein and energy levels on the protein and lipid content of the hen's egg. *Poultry Science* 51:994–997.

Garret, M.G. and W.L. Franklin. 1988. Behavioral ecology of dispersal in the black-tailed prairie dog. *Journal of Mammalogy* 69:236–250.

Gates, J.E. and L.W. Gysel. 1978. Avian nest dispersion and fledging success in field-forest ecotones. *Ecology* 59:871–883.

Giuliano, W.M. and S.E. Daves. 2002. Avian response to warm-season grass use in pasture and hayfield management. *Biological Conservation* 106:1–9.

Godfrey, G.A. 1967. Summer and fall movements and behavior of immature ruffed grouse. MS Thesis, University of Minnesota, Minneapolis, MN, USA.

———. 1975a. Underestimation experienced in determining ruffed grouse brood size. *Journal of Wildlife Management* 39:191–193.

———. 1975b. Home range characteristics of ruffed grouse broods in Minnesota. *Journal of Wildlife Management* 39:287–298.

——— and W.H. Marshall. 1969. Brood break-up and dispersal of ruffed grouse. *Journal of Wildlife Management* 33:609–620.

Gordon, D.S. 2005. Third year effects of shelterwood cutting, wildlife retention cutting, and prescribed burning on oak regeneration, understory vegetation development, and acorn production in Tennessee. MS Thesis, University of Tennessee. Knoxville, TN, USA.

Gormley, A. 1996. Causes of mortality and factors affecting survival of ruffed grouse in Northern Michigan. MS Thesis, Michigan State University, East Lansing, MI, USA.

Grange, W.B. 1948. Wisconsin Grouse Problems. Wisconsin Conservation Department, Madison, Wisconsin.

Gratson, M.W. and C.L. Whitman. 2000. Characteristics of Idaho elk hunters relative to road access on public lands. Wildlife Society Bulletin 28:1016–1022.

Greenberg, C.H., D.J. Levey, and D.L. Loftis. 2007. Fruit production in mature and recently regenerated upland and cove hardwood forests of the southern Appalachians. *Journal of Wildlife Management.* 71:321–335.

Greenwood, P.J. 1980. Mating systems, philopatry and dispersal in birds and mammals. *Animal Behavior* 28:1140–1162.

———. and P.H. Harvey. 1982. The natal and breeding dispersal of birds. *Annual Review of Ecology and Systematics* 13:1–21.

———, ———, and C.M. Perrins. 1979. The role of dispersal in the great tit *(Parus major):* the causes, consequences and heritability of natal dispersal. *Journal of Animal Ecology* 48:123–142.

Gullion, G.W. 1965. Improvements in methods for trapping and marking ruffed grouse. *Journal of Wildlife Management* 29:109–116.

———. 1970. Factors affecting ruffed grouse populations in the boreal forests of Northern Minnesota, USA. International Congress of Game Biologists 8:103–117.

———. 1977. Forest manipulation for ruffed grouse. Transaction of the North American Natural Resources Conference 42:449–458.

———. 1981a. The impact of goshawk predation upon ruffed grouse. *Loon* 53:82–84.

———. 1981b. Non-drumming males in a ruffed grouse population. Wilson Bulletin 93:372–382.

———. 1983. Ruffed grouse habitat manipulation: Mille Lacs Wildlife Management Area, Minnesota. *Minnesota Wildlife Research Quarterly* 43:25–98.

———. 1984a. Managing northern forests for wildlife. Minnesota Agricultural Experiment Station, Miscellaneous Publication 13,442.

———. 1990. Ruffed grouse use of conifer plantations. Wildlife Society Bulletin 18:183–187.

———, and W.H. Marshall. 1968. Survival of ruffed grouse in a boreal forest. *Living Bird* 7:117–167.

Gutierrez, R.J., G.S. Zimmerman, and G.W. Gullion. 2003. Daily survival rates of ruffed grouse *(Bonasa umbellus)* in northern Minnesota. *Wildlife Biology* 9:351–356.

Guyette, R.P., R.M. Muzika, J. Kabrick, and M.C. Stambaugh. 2004. A perspective on *Quercus* life history characteristics and forest disturbance, *in* M.A. Spetich, editor. Proceedings of upland oak ecology symposium: history, current conditions, and sustainability. U.S. Forest Service Southern Research Station General Technical Report SRS– 73. Asheville, North Carolina, USA.

Hale, P.E., A.S. Johnson, and J.L. Landers. 1982. Characteristics of ruffed grouse drumming sites in Georgia. Journal of Wildlife Management 46:115-123.

Hall, L.S., P.R. Krausmann, and M.L. Morrison. 1997. The habitat concept and a plea for standard terminology. Wildlife Society Bulletin 25:173–182.

Harper, C.A. 2008. A guide to successful wildlife food plots: Blending science with common sense. UT Extension, PB 1769. Knoxville, TN, 168 pages.

———, G.E. Bates, M.P. Hansbrough, M.J. Gudlin, J.P. Gruchy, and P.D. Keyser. 2007. Native warm-season grasses: Identification, establishment, and management for wildlife and forage production in the Mid-South. UT Extension, PB 1752, Knoxville, TN. 189 pp.

———, C.E. Moorman, and P.D. Keyser. *2008*. Native grass and early successional wildlife habitat: Past lessons and a new vision. Proceedings Eastern Native Grass Symposium 6:120-126.

Harris, M.J. 1981. Spring and summer ecology of ruffed grouse in northern Georgia. MS Thesis, University of Georgia, Athens, GA, USA.

Haulton, G.S. 1999. Ruffed grouse natality, chick survival, and brood micro-habitat selection in the southern Appalachians. MS Thesis, Virginia Tech, Blacksburg, VA, USA.

———, D.F. Stauffer, R.L. Kirkpatrick, and G.W. Norman. 2003. Ruffed grouse *(Bonasa umbellus)* brood microhabitat selection in the southern Appalachians. *American Midland Naturalist* 150:95–103.

Hayes, S.G., D.J. Deptich, and P. Zager. 2002. Proximate factors affecting elk hunting in northern Idaho. *Journal of Wildlife Management* 66:491–499.

Healy, W.M. 1977. Wild turkey winter habitat in West Virginia cherry-maple forests. *Transactions of the Northeast Section of The Wildlife Society* 34:7–12.

——— and E.S. Nenno. 1983. Minimum maintenance versus intensive management of clearings for wild turkeys. Wildlife Society Bulletin 11:113–120.

——— and J.C. Pack. 1983. Managing seeps for wild turkeys in northern hardwood forest types in West Virginia. *Transactions of the Northeast Section of The Wildlife Society* 40:19–30.

Heard, L.P., A.W. Allen, L.B. Best, S.J. Brady, L.W. Burger, A.J. Esser, E. Hackett, D.H. Johnson, R.L. Pederson, R.E. Reynolds, C. Rewa, M.R. Ryan, R.T. Molleur, and P. Buck. 2000. A comprehensive review of Farm Bill contributions to wildlife conservation, 1985–2000. W.L. Holman and D.J. Halloum, editors. US Department of Agriculture, Natural Resources Conservation Service, Wildlife Habitat Management Institute, Technical Report, USDA/NRCS/WHMI-2000.

Hewitt, D.G. 1994. Ruffed grouse nutrition and foraging in the southern Appalachians. PhD Dissertation, Virginia Tech, Blacksburg, VA, USA.

——— and R.L. Kirkpatrick. 1996. Forage intake rates of ruffed grouse and potential effects on grouse density. *Canadian Journal of Zoology* 74:2016–2024.

——— and ———. 1997*a*. Ruffed grouse consumption and detoxification of evergreen leaves. Journal of Wildlife Management 61:129–139.

——— and ———. 1997*b*. Daily activity times of ruffed grouse in southwestern Virginia. *Journal of Field Ornithology* 68:413–420.

Hines, J.E. 1986. Survival and reproduction of dispersing blue grouse. *Condor* 88:43–49.

Hoffman, R.W. 1991. Spring movements, roosting activities and home range characteristics of male Merriam's wild turkey. *Southwestern Naturalist* 36:332–337.

Hollifield, B.K. and R.W. Dimmick. 1995. Arthropod abundance relative to forest management practices benefiting ruffed grouse in the southern Appalachians. Wildlife Society Bulletin 23:756–764.

Holmes, R.T. and T. W. Sherry. 2001. Thirty-year bird population trends in an unfragmented temperate deciduous forest: Importance of habitat change. *Auk.* 118:589–609.

Hopkins, A.D. 1938. Bioclimatics: a science of life and climate relations. U.S. Department of Agriculture Miscellaneous Publication 280. Washington, DC, USA.

Hunter, W.C., D.A. Buehler, R.A. Canterbury, J.L. Confer, and P.B. Hamel. 2001. Conservation of disturbance dependent birds in eastern North America. Wildlife Society Bulletin 29:440–455.

Jackson, S.W. 2002. First-year changes in oak regeneration, understory competitors, and resource levels in response to two overstory treatments and prescribed burning at Chuck Swan State Forest. Thesis, University of Tennessee. Knoxville, Tennessee, USA.

———, R.G. Basinger, D.S. Gordon, C.A. Harper, D.S. Buckley, and D.A. Buehler. 2006. Influence of silvicultural treatments on eastern wild turkey habitat characteristics in eastern Tennessee. Proceedings National Wild Turkey Symposium 9:199-207.

Johnsgard, P.A. and S.J. Maxson. 1989. Nesting. Pages 130–137 *in* S. Atwater and J. Schnell, editors. *The wildlife series: ruffed grouse.* Stackpole Books, Harrisburg, Pennsylvania, USA.

———, R.O. Kimmel, S.J. Maxson, J.G. Scott, R. Small, and G.L. Storm. 1989. The young grouse. Pages 140–159 *in* S. Atwater and J. Schnell, editors. *The wildlife series: Ruffed Grouse.* Stackpole Books, Harrisburg, Pennsylvania, USA.

Johnson, A.S. and P.E. Hale. 2002. The historical foundations of prescribed burning for wildlife: A Southern perspective. *In* W.M. Ford, K.R. Russell, and C.E. Moorman, editors. The Role of Fire in Nongame Wildlife Management and Community Restoration: Traditional Uses and New Directions. USDA Forest Service General Technical Report NE-288.

Johnson, D.H. 1980. The comparison of usage and availability measurements for evaluating resource preference. *Ecology* 61:65–71.

Johnson, M.L. and M.S. Gaines. 1987. The selective basis for dispersal of the prairie vole, *Microtus ochrogaster. Ecology* 68:684–694.

Jones, B.C. 2005. Ruffed grouse habitat use, reproductive ecology, and survival in western North Carolina. PhD Dissertation, University of Tennessee. Knoxville, TN, USA.

——— and C.A. Harper. 2006. Ruffed grouse *(Bonasa umbellus)* use of stands harvested via alternative regeneration techniques in the southern Appalachians. 15th Central Hardwoods Forest Conference. Knoxville, Tennessee, USA.

———, ———, D.A. Buehler, and G.S. Warburton. 2005. Use of spring drumming counts to index ruffed grouse populations in the southern Appalachians. Proceedings Annual Conference of Southeastern Association of Fish and Wildlife Agencies 59:135-143.

Jones, B.C., J.F. Kleitch, C.A. Harper, and D.A. Buehler. 2008. Ruffed grouse brood habitat use in a mixed hardwood forest: Implications for forest management in the Appalachians. *Forest Ecology and Management* 255:3580–3588.

Kalla, P.I. and R.W. Dimmick. 1995. Reliability of established aging and sexing methods in ruffed grouse. Proceeding of the Annual Conference of the Southeastern Association of Fish and Wildlife Agencies. 49:580–593.

Keith, L.B., A.W. Todd, C.J. Brand, R.S. Adamcik, and D.H. Rusch. 1977. An analysis of predation during a cyclic fluctuation of snowshoe hares. Proceedings of the International Congress of Game Biologists 13:151–175.

Kenward, R.E. 2001. *A manual for wildlife radio tagging.* Academic Press, New York, NY.

Keppie, D.M. 2004. Autumn dispersal and winter residency do not confer reproductive advantages on female Spruce Grouse. *Condor* 106:896–904.

Kernohan, B.J., R.A. Gitzen, and J.J. Millspaugh. 2001. Analysis of animal space use and movements, pp. 125–166. *In* J.J. Millspaugh and J.M. Marzluff [Eds.], *Radio tracking and animal populations.* Academic Press, New York, NY.

Kilgo, J.C., R.F. Labisky, and D.E. Fritzen 1998. Influences of hunting on the behavior of white-tailed deer: implications for conservation of the Florida panther. *Conservation Biology* 12:1359–1364.

Kimmel, R.O. and D.E. Samuel. 1978. Feeding behavior of young ruffed grouse in West Virginia. *Transactions of the Northeast Section of the Wildlife Society* 35:43–51.

Knight, R.L. and D.N. Cole. 1995. Wildlife responses to recreationists. Pages 51–70 *in* R. L. Knight and K. J. Gutzwiller, editors. *Wildlife and recreationists: coexistence through management and research.* Island Press, Washington DC, USA.

Koenig, W.D. and J.M.H. Knops. 2002. The behavioral ecology of masting in oaks. Pp 129–148 *in* W.J. McShea and W.M. Healy, Eds. *Oak forest ecosystems: ecology and management for wildlife.* Johns Hopkins University Press, Baltimore, MD.

Kubisiak, J.F. 1978. Brood characteristics and summer habitats of ruffed grouse in central Wisconsin. Wisconsin Department of Natural Resources Technical Bulletin 108, Madison, Wisconsin, USA.

———. 1984. The impact of hunting on ruffed grouse populations in the Sandhill Wildlife Area. Pages 151–168 *in* William L. Robinson, editor. *Ruffed grouse management: state of the art in the early 1980's.* The North Central Section of The Wildlife Society and the Ruffed Grouse Society.

———, J.C. Moulton, and K.R. McCafferty. 1980. Ruffed grouse density and habitat relationships in Wisconsin. Department of Natural Resources Technical Bulletin No. 118, Madison, Wisconsin, USA.

Lack, D. 1954. *The natural regulation of animal numbers.* Clarendon Press, Oxford. 343 pp.

Lambin, X., J. Aars, and S.B. Piertney. 2001. Dispersal, intraspecific competition, kin competition, and kin facilitation: a review of the empirical evidence, pp. 110–122. *In* J. Clobert, E. Danchin, A.A. Dhondt, and J.D. Nichols (eds.), *Dispersal*. Oxford University Press, NY.

Larson, M.A., M.E. Clark, and S.R. Winterstein. 2001. Survival of Ruffed Grouse chicks in northern Michigan. Journal of Wildlife Management 65:880–886.

———, ———, and ———. 2003. Survival and habitat of ruffed grouse nests in northern Michigan. Wilson Bulletin 115:140–147.

Lawson, E.J.G. and A.R. Rodgers 1997. Differences in home range size computed in commonly used software programs. Wildlife Society Bulletin 25:721–729.

Lenth, R.V. 1981. On finding the source of a signal. *Technometrics* 23:149–154.

Litvaitis, J.A. 2001. Importance of early successional habitats to mammals in eastern forests. Wildlife Society Bulletin 29:466–473.

Loftis, D.L. 1983. Regenerating southern Appalachian mixed hardwoods with the shelterwood method. *Southern Journal of Applied Forestry* 7:212–217.

———. 1990. A shelterwood method for regenerating red oak in the southern Appalachians. *Forest Science* 36:917–929.

———. 1993. Regenerating northern red oak on high-quality sites in the southern Appalachians. *In* D.L. Loftis and C.E. McGee, editors, Oak regeneration: Serious problems, practical recommendations. Symposium proceedings, Knoxville, Tennessee. General Technical Report SE-84. US Department of Agriculture, Forest Service, Southeastern Forest Experiment Station.

Long, C.R. 2007. Pre-breeding food habitats and condition of ruffed grouse and effects on reproduction in the central and southern Appalachians. MS Thesis, West Virginia University, Morgantown, WV, USA.

——— and J. Edwards. 2004a. Pre-breeding nutritional condition and potential effects on reproduction. *In* G.W. Norman, D.F. Stauffer, J. Sole, T.J. Allen, W.I. Igo, S. Bittner, J. Edwards, R.L. Kirkpatrick, W.M. Giuliano, B. Tefft, C.A. Harper, D.A. Buehler, D. Figert, M. Seamster, D. Swanson, editors, Ruffed grouse ecology and management in the Appalachian region. Final Project Report of the Appalachian Cooperative Grouse Research Project.

——— and ———. 2004b. Pre-breeding food habits of ruffed grouse in the Appalachian region. *In* G.W. Norman, D.F. Stauffer, J. Sole, T.J. Allen, W.I. Igo, S. Bittner, J. Edwards, R.L. Kirkpatrick, W.M. Giuliano, B. Tefft, C.A. Harper, D.A. Buehler, D. Figert, M. Seamster, D. Swanson, editors, Ruffed grouse ecology and management in the Appalachian region. Final Project Report of the Appalachian Cooperative Grouse Research Project.

Lorimer, C.G. 1993. Causes of the oak regeneration problem. *In* D.L. Loftis and C.E. McGee, editors, Oak regeneration: Serious problems, practical recommendations. Symposium proceedings, Knoxville, Tennessee. General Technical Report SE-84. US Department of Agriculture, Forest Service, Southeastern Forest Experiment Station.

———. 2001. Historical and ecological roles of disturbance in eastern North American forests: 9,000 years of change. Wildlife Society Bulletin 29:425–439.

Lovallo, M.J., D.S. Klute, G.L. Storm, and W.M. Tzilkowski 2000. Alternate drumming site use by ruffed grouse in Pennsylvania. *Journal of Field Ornithology* 71:506–515.

Lyon, L.J., and M.G. Burcham 1998. Tracking elk hunters with the Global Positioning System. Research Paper, USDA Forest Service, Rocky Mountain Research Station, Ogden, UT. RMRS-RP-3.

Macdonald, J.E., W.L. Palmer, and G.L. Storm. 1994. Ruffed grouse population response to intensive forest management in central Pennsylvania, USA. *In* I.D. Thompson, editor, Proceedings of the XXI International Union of Game Biologists Congress, Volume 2.

Marshall, W.H. 1965. Ruffed grouse behaviour. *Bioscience* 15:92–94.

——— and G.W. Gullion. 1965. A discussion of ruffed grouse populations. Cloquet Forest Research Center, Minnesota. Transactions of International Union Game Biologists 6:93–102.

Maxson, S. 1977. Activity patterns of female ruffed grouse during the breeding season. Wilson Bulletin 89:439–455.

———. 1978. Spring home range and habitat use by female ruffed grouse. *Journal of Wildlife Management* 42:61–71.

Mayfield, H. 1961. Nesting success calculated from exposure. Wilson Bulletin 73:255–261.

McCaffery, K.R., J.E. Ashbrenner, W.A. Creed, and B.E. Kohn. 1996. Integrating forest and ruffed grouse management: A case study at the Stone Lake Area. Wisconsin Department of Natural Resources, Technical Bulletin 189. Madison, WI. 39 pages.

McCorquodale, S.M., R. Wiseman, and C.L. Marcum. 2003.

Survival and harvest vulnerability of elk in the Cascade Range of Washington. *Journal of Wildlife Management* 67:248–257.

Menge, H., T. Frobish, B.T. Weinland, and E.G. Geis. 1979. Effect of dietary protein and energy on reproductive performance of turkey hens. *Poultry Science* 58:419–426.

Miller, G.W. and T.M. Schuler. 1995. Development and quality of reproduction in two-age central Appalachian hardwoods — 10-year results. Proceedings of the 10th Central Hardwood Conference, USDA Forest Service, General Technical Report NE-197. K.W. Gottschalk and S.L.C. Fosbroke, editors. Morgantown, West Virginia, USA.

Monschein, T.D. 1974. Effects of hunting on ruffed grouse in small woodlots in Ashe and Alleghany Counties, North Carolina. Proceedings of the 27th Annual Conference of the Southeast Association of Game and Fish Commissioners. 27:30–36.

Mueller, H.C. and D.D. Berger. 1992. Behavior of migrating raptors: differences between spring and fall. *Journal of Raptor Research* 26:136–145.

Neher, L.N. 1993. Winter movements and habitat selection of ruffed grouse in central Missouri. Thesis, University of Missouri, Columbia, Missouri. 69 pp.

Nilsson, J.A. 1989. Establishment of juvenile marsh tits in winter flocks: an experimental study. *Animal Behavior* 38:586–595.

Norman, G.W. 2006. 2006-06 Ruffed grouse population status in Virginia. Wildlife Resource Bulletin No. 06-2. Department of Game and Inland Fisheries, Richmond, VA, USA, 21pp.

——— and R.L. Kirkpatrick. 1984. Foods, nutrition, and condition of ruffed grouse in southwestern Virginia. *Journal of Wildlife Management* 48:183–187.

———, D.F. Stauffer, J.D. Sole, T.J. Allen, W.K. Igo, S. Bittner, J.W. Edwards, R.L. Kirkpatrick, W.M. Giuliano, B. Tefft, C. Harper, D. Buehler, D.E. Figert, M. Seamster, and D. Swanson. 2004. Ruffed Grouse ecology and management in the Appalachian region. Final Project Report of the Appalachian Cooperative Grouse Research Project.

Ott, R.A. 1990. Winter roosting energetics, roost site selection and time activity budgets of ruffed grouse in central New York. Thesis, State University of New York, College of Environmental Science and Forestry, Syracuse, New York, USA.

Pack, J.C., K.I. Williams, and C.I. Taylor. 1988. Use of prescribed burning to increase wild turkey brood range habitat in oak-hickory forests. *Transactions of the Northeast Section of The Wildlife Society* 45:37–48.

Pagen, R.W., F.R. Thompson III, and D.E. Burhans. 2000. Breeding and post-breeding habitat use by forest migrant songbirds in the Missouri Ozarks. *Condor* 102:738–747.

Palmer, W.L. 1956. Ruffed grouse population studies on hunted and unhunted areas. Twenty-first North American Wildlife Conference.

Palmer, W.L. and C.L. Bennett, Jr. 1963. Relation of season length to hunting harvest of ruffed grouse. *Journal of Wildlife Management* 27:634–639.

Pedersen, H.C., S.L. Kastdalen, H. Broseth, R.A. Ims, W. Svendesen, and N.G. Yoccoz. 2004. Weak compensation of harvest surplus despite strong density dependent growth in willow ptarmigan. Proceedings of the Royal Society of London — Biological Sciences 271:381–385.

Plaugher, G.F. 1998. Seasonal habitat, foods, and movements of ruffed grouse in the central Appalachian Mountains of West Virginia. MS Thesis, West Virginia University, Morgantown, WV, USA.

Powell, R.A. 2000. Animal home ranges and territories and home range estimators. Pages 65–110 *in* L. Boitani and T.K. Fuller [Eds.]. *Research techniques in animal ecology: controversies and consequences.* Columbia University Press, New York, NY.

Proctor, A.B. 2010. Effect of nutritional deficiency on ruffed grouse condition and reproductive success. MS Thesis, West Virginia University, Morgantown, WV, USA.

Reynolds, M.C., D.F. Stauffer, R.L. Kirkpatrick, and G.W. Norman. 2000. A summary of findings from Phase I of the research project. Appalachian Cooperative Grouse Research Project, VA Department of Game and Inland Fisheries, Verona, VA.

Robbins, C.T. 1981. Estimation of the relative cost of reproduction in birds. *Condor* 83:177–179.

Rodewald, A.D. and R.H. Yahner. 2001. Avian nesting success in forested landscapes: influence of landscape composition, stand and nest-patch microhabitat, and biotic interactions. *Auk* 118:1018–1028.

Rogers, R.E. and D.E. Samuel. 1984. Ruffed grouse brood use of oak-hickory managed with prescribed burning. *Transactions of the Northeast Section of The Wildlife Society* 41:142–154.

Rusch, D.H., S. DeStephano, M.C. Reynolds, and D. Lauten. 2000. Ruffed grouse (Bonasa umbellus). Number 515 in A. Poole and F. Gill, editors. The birds of North America. The Birds of North America, Philadelphia, Pennsylvania, USA.

——— and P.D. Doerr. 1972. Broad-winged hawk nesting and food habits. *Auk.* 89:139–145.

——— and L.B. Keith. 1971. Seasonal and annual trends in Alberta ruffed grouse. *Journal of Wildlife Management* 35:803–822.

———, E.C. Meslow, L.B. Keith, and P.D. Doerr. 1972. Response of great horned owl populations to changing prey densities. *Journal of Wildlife Management.* 36:282–296.

SAMAB - Southern Appalachian Man and the Biosphere Cooperative. 1996. The Southern Appalachian Assessment Terrestrial Technical Report. Report 5 of 5. Atlanta. U.S. Dept of Agriculture, Forest Service, Southern Region.

Sander, I.L., C.E. McGee, K.G. Day, and R.E. Willard. 1983. Oak-Hickory. *In* R.M. Burns, editor. Silvicultural systems for the major forest types of the United States. US Department of Agriculture, Forest Service, Agriculture Handbook 445.

Sauer, J.R., J.E. Hines, and J. Fallon. 2004. The North American Breeding Bird Survey, Results and Analysis 1966–2003. Version 2004.1. USGS Patuxent Wildlife Research Center, Laurel, Maryland, USA.

Schumacher, C.L. 2002. Ruffed grouse habitat use in western North Carolina. MS Thesis, University of Tennessee. Knoxville, TN, USA.

———, C.A. Harper, D.A. Buehler, G.S. Warburton, and W.G. Minser, III. 2001. Drumming log habitat selection by male ruffed grouse in North Carolina. Proceedings Annual Conference of Southeastern Association of Fish and Wildlife Agencies 55:466–474.

Scott, J.G., M.J. Lovallo, G.L. Storm, and W.M. Tzilkowski. 1998. Summer habitat use by ruffed grouse broods in central Pennsylvania. *Journal of Field Ornithology* 69:474–485.

Seaman, D.E., J.J. Millspaugh, B.J. Kernohan, G.C. Brundige, K.J. Raedeke, and R.A. Gitzen 1999. Effects of sample size on kernel home range estimates. *Journal of Wildlife Management* 63:739–747.

——— and R.A. Powell 1996. An evaluation of the accuracy of kernel density estimators for home range analysis. *Ecology* 77:2075–2085.

Seehorn, M.E., R.F. Harlow, and M.T. Mengak. 1981. Foods of ruffed grouse from three locations in the southern Appalachian mountains. Proceedings of the Annual Conference of the Southeast Association of Fish and Wildlife Agencies 35:216–224.

Servello, F.A. and R.L. Kirkpatrick. 1986. Sexing ruffed grouse in the Southeast using feather criteria. Wildlife Society Bulletin 14:280–282.

——— and ———. 1987. Regional variation in the nutritional ecology of ruffed grouse. *Journal of Wildlife Management* 51:749–770.

——— and ———. 1988. Nutrition and condition of ruffed grouse during the breeding season in southwestern Virginia. *Condor* 90:836–842.

——— and ———. 1989. Nutritional value of acorns for ruffed grouse. *Journal of Wildlife Management* 53:26–29.

Sharp, W.M. 1963. The effect of habitat manipulation and forest succession on ruffed grouse. *Journal of Wildlife Management* 27:664–671.

Siepielski, A.M., A.D. Rodewald, and R.H. Yahner. 2001. Nest site selection and nesting success of the red-eyed vireo in central Pennsylvania. Wilson Bulletin 113:302–307.

Small, R.J. 1985. Mortality and dispersal of ruffed grouse in central Wisconsin. Thesis, University of Wisconsin, Madison, WI. 56 pp.

———, J.C. Holzwart, and D.H. Rusch. 1991. Predation and hunting mortality of ruffed grouse in central Wisconsin. *Journal of Wildlife Management* 55:512–520.

———, ———, and ———. 1993. Are ruffed grouse more vulnerable to mortality during dispersal? *Ecology* 74:2020–2026.

———, ———, and ———. 1996. Natality of ruffed grouse Bonasa umbellus in central Wisconsin, USA. Wildlife Biology 2:49–52.

———, and D.H. Rusch. 1989. The natal dispersal of ruffed grouse. *Auk* 106:72–79.

Smith, A.F. 1977. Fall and winter food habits of the ruffed grouse in Georgia. MS Thesis, University of Georgia, Athens, GA, USA

Smith, A.T. 1974. The distribution and dispersal of pikas: influences of behavior and climate. *Ecology* 55:1368–1376.

Smith, B.W. 2006. Nesting Ecology, Chick Survival, and Juvenile Dispersal of ruffed grouse *(Bonasa umbellus)* in the Appalachian Mountains. PhD Dissertation, West Virginia University, Morgantown, WV, USA.

———, C. Dobony, and J. Edwards. 2004. Nesting and reproductive output: survival and cause specific mortality of ruffed grouse chicks in the Appalachians. Pages 27–28 in G.W. Norman, D.F. Stauffer, J. Sole, T.J. Allen, W.K. Igo, S. Bittner, J. Edwards, R.L. Kirkpatrick, W.M. Giuliano, B. Tefft, C. Harper, D. Buehler, D. Figert, M. Seamster, and D. Swanson, editors. Ruffed grouse ecology and management in the Appalachian region. Final Project Report of the Appalachian Cooperative Grouse Research Project.

Smith, D.M. 1986. *The practice of silviculture.* John Wiley and Sons, New York, USA.

Smith, D.W. 1995. The southern Appalachian hardwood region. Pages 173–226 *in* J.W. Garrett, editor. *Regional silviculture of the United States.* Third edition. John Wiley & Sons, New York, New York, USA.

Smith, H.C., L. Della-Bianca, and H. Fleming. 1983. Appalachian mixed hardwoods. *In* R.M. Burns, editor. Silvicultural systems for the major forest types of the United States. US Department of Agriculture, Forest Service, Agriculture Handbook 445.

———, N.I. Lamson, and G.W. Miller. 1989. An esthetic alternative to clearcutting? *Journal of Forestry* 87:14–18.

Stafford, S.K. and R.W. Dimmick. 1979. Autumn and winter foods of ruffed grouse in the southern Appalachians. *Journal of Wildlife Management* 43:121–127.

Stewart, R.E. 1956. Ecological study of ruffed grouse broods in Virginia. *Auk* 73:33–41.

Stoll, R.J., W.L. Culbertson, M.W. McClain, R.W. Donohoe, and G. Honchul. 1999. Effects of clearcutting on ruffed grouse in Ohio's oak-hickory forests. Ohio Department of Natural Resources Division of Wildlife Report 7, Columbus, Ohio, USA.

Stollberg, B.P., R.L. Hine, J.B. Hale, Eds. 1952. Food habit studies of ruffed grouse, pheasant, quail and mink in Wisconsin. Wisconsin Department of Natural Resources Technical Bulletin 4. 22 pp.

Storm, G.L., W.L. Palmer, and D.R. Diefenbach. 2003. Ruffed grouse response to management of mixed oak and aspen communities in central Pennsylvania. Pennsylvania Game Commission Grouse Research Bulletin Number 1. Harrisburg, Pennsylvania, USA.

Strickland, M.D., H.J. Harju, K.R. McCaffery, H.W. Miller, L.M. Smith, and R.J. Stoll. 1994. Harvest Management. Pages 445 – 473 *in* T. A. Bookhout, editor. *Research and management techniques for wildlife and habitats.* Fifth edition. The Wildlife Society, Bethesda, Maryland.

Svoboda, F.J. and G.W. Gullion. 1972. Preferential use of aspen by ruffed grouse in northern Minnesota. *Journal of Wildlife Management* 36:1166–1180.

Swank, W.T., J.M. Vose, and K.J. Elliott. 2001. Long-term hydrologic and water quality response following commercial clearcutting of mixed hardwoods on a southern Appalachian catchment. *Forest Ecology and Management,* 26pp.

Swanson, D.A., M.C. Reynolds, J.M. Yoder, R.J. Stoll, Jr., and W.L. Culbertson. 2003. Ruffed grouse survival in Ohio. Ohio Wildlife Research Report.

Swift, L.W., Jr. 1984. Soil losses from roadbeds and cut and fill slopes in the Southern Appalachian Mountains. *Southern Journal of Applied Forestry* 8:209–215.

———. 1985. Forest road design to minimize erosion in the southern Appalachians. *In* B.G. Blackmon, editor. Proceedings of Forestry and Water Quality: A Mid-South Symposium. Little Rock, Arkansas.

———. 1988. Forest access roads: Design, maintenance, and soil loss. *In* W.T. Swank and D.A. Crossley Jr., editors. *Ecological Studies, Volume 66: Forest Hydrology and Ecology at Coweeta.* Springer-Verlag, New York.

Thomas, V.G., H.G. Lumsden, and D.H. Price. 1975. Aspects of the winter metabolism of ruffed grouse *(Bonasa umbellus)* with special reference to energy reserves. *Canadian Journal of Zoology* 53:434–440.

Thompson, F.R. III. 1987. The ecology of ruffed grouse in central Missouri. PhD Dissertation, University of Missouri, Columbia, MO, USA

——— and D.R. Dessecker. 1997. Management of early-successional communities in central hardwood forests: with special emphasis on the ecology and management of oaks, ruffed grouse, and forest songbirds. Gen. Tech. Rep. NC-195. St. Paul, MN: U.S. Department of Agriculture, Forest Service, North Central Forest Experiment Station. 33 pp.

———, W.D. Dijak, T.G. Kulowiec, and D.A. Hamilton. 1992. Breeding bird populations in Missouri Ozark forests with and without clearcutting. *Journal of Wildlife Management.* 56:23–30.

———, D.A. Freiling, and E.K. Fritzell. 1987. Drumming, nesting, and brood habitats of ruffed grouse in an oak-hickory forest. *Journal of Wildlife Management* 51:568–575.

——— and E.K. Fritzell. 1988a. Ruffed grouse metabolic rate and temperature cycles. *Journal of Wildlife Management* 52:450–453.

——— and ———. 1988*b*. Ruffed Grouse winter roost site preference and influence on energy demands. *Journal of Wildlife Management* 52:454–460.

——— and ———. 1989. Habitat use, homerange, and survival of territorial male ruffed grouse. *Journal of Wildlife Management* 53:15–21.

Tirpak, J.M. 2000. Influence of microhabitat structure on nest success and brood survival of ruffed grouse in the central and southern Appalachians. MS Thesis, California University of Pennsylvania, California, PA, USA.

———. 2005. Modeling ruffed grouse populations in the central and southern Appalachians. PhD Dissertation, Fordham University, Bronx, New York, USA.

———, W.M. Giuliano, and C.A. Miller. 2005. Nocturnal roost habitat selection by ruffed grouse broods. *Journal of Field Ornithology* 76:168–174.

———, ———, ———, T.J. Allen, S. Bittner, J.W. Edwards, S. Friedhoff, W.K. Igo, D.F. Stauffer, and G.W. Norman. 2006. Ruffed grouse nest success and habitat selection in the central and southern Appalachians. *Journal of Wildlife Management* 70:138–144.

Trefethen, J.B. 1975. An American crusade for wildlife. The Boone and Crockett Club, Alexandria, VA USA.

Triquet, A.M. 1989. Mortality and habitat use of ruffed grouse on the Cumberland Plateau, Kentucky. PhD Dissertation, University of Kentucky, Lexington, KY, USA.

Tubbs, C.H., R.D. Jacobs, and D. Cutler. 1983. Northern Hardwoods. *In* R.M. Burns, editor. Silvicultural systems for the major forest types of the United States. US Department of Agriculture, Forest Service, Agriculture Handbook 445.

USDA Forest Service. 2002. Forest Inventory and Analysis Data Base: Forest Inventory Mapmaker Version 1.7. U.S. Dept of Agriculture.

Van Lear, D.H. and P.H. Brose. 2002. Fire and oak management. *In* W.J. McShea and W.M. Healy, editors. *Oak forest ecosystems: Ecology and management for wildlife.* Johns Hopkins University Press. Baltimore, Maryland, USA.

——— and R.F. Harlow. 2002. Fire in the eastern United

States: Influence on wildlife habitat. *In* W.M. Ford, K.R. Russell, and C.E. Moorman, editors. The Role of Fire in Nongame Wildlife Management and Community Restoration: Traditional Uses and New Directions. USDA Forest Service General Technical Report NE-288.

——— and T.A. Waldrop. 1989. History, uses, and effects of fire in the Appalachians. USDA Forest Service General Technical Report SE-54.

Vega Rivera, J.H., J.H. Rappole, W.J. McShea, and C.A. Haas. 1998. Wood thrush post fledgling movements and habitat use in northern Virginia. *Condor* 100:69–78.

Washburn, B.E., T.G. Barnes, and J.D. Sole. 2000. Improving northern bobwhite habitat by converting tall fescue fields to native warm-season grasses. Wildlife Society Bulletin 28:97–104.

Whitaker, D.M. 2003. Ruffed grouse *(Bonasa umbellus)* habitat ecology in the central and southern Appalachians. Ph.D. dissertation, Virginia Tech, Blacksburg, VA, USA.

——— and D.F. Stauffer. 2003. Night roost selection during winter by ruffed grouse in the central Appalachians. *Southeastern Naturalist* 2:377–392.

———, ———, G.W. Norman, and B. Chandler. 2004. Effect of prescribed burning of clearcuts on ruffed grouse brood habitat. Proceedings of the Annual Conference of the Southeastern Association of Fish and Wildlife Agencies 58:312–322.

———, ———, ———, P.K. Devers, T.J. Allen, S. Bittner, D. Buehler, J. Edwards, S. Friedhoff, W.M. Giuliano, C.A. Harper, B. Tefft. 2006. Factors affecting habitat us by Appalachian ruffed grouse. *Journal of Wildlife Management* 70:460–471.

———, ———, ———, ———, J. Edwards, W.M. Giuliano, C.A. Harper, W. Igo, J. Sole, H. Spiker, and B. Tefft. 2007. Factors associated with variation in home range size of Appalachian ruffed grouse. *Auk*. 124:1407–1424.

White, D.W. and R.W. Dimmick 1979. The Distribution of ruffed grouse in Tennessee. *Journal of the Tennessee Academy of Science* 54:114–115.

White, G.C. and R.A. Garrott. 1990. *Analysis of wildlife radio-tracking data.* Academic Press, New York, NY, USA.

Wiggers, E.P., M.K. Laubhan, and D.A. Hamilton 1992. Forest structure associated with ruffed grouse abundance. *Forest Ecology and Management* 49:211–218.

Wiggett, D.R. and D.A. Boag. 1989. Intercolony natal dispersal in the Columbian ground squirrel. *Canadian Journal of Zoology* 67:42–50.

Williams, T.D. 1994. Intraspecific variation in egg size and egg composition in birds: Effects on offspring fitness. Biological Reviews of the Cambridge Philosophical Society 68:35–59.

Woehr, J.R. and R.E. Chambers. 1975. Winter and spring food preference of ruffed grouse in central New York. Transactions *of the Northeast Section of the Wildlife Society* 32:95–110.

Worton, B.J. 1989. Kernel methods for estimating the utilization distribution in home range studies. Ecology 70:164–168.

Wright, H.A. and A.W. Bailey. 1982. *Fire ecology: United States and southern Canada.* John Wiley and Sons, New York, USA.

Wunz, G.A., A.H. Hayden, and R.R. Potts, III. 1983. Spring seep ecology and management. *Transactions of the Northeast Section of The Wildlife Society* 40:19–30.

Yahner, R.H., C.G. Mahan, and C.A. Delong. 1993. Dynamics of depredation on artificial nests in habitat managed for ruffed grouse. Wilson Bulletin 105:172–179.

Yoder, J.M. 2004. Ruffed grouse dispersal: relationships with landscape and consequences for survival. Ph.D. dissertation, The Ohio State University, Columbus, OH, USA.

———, E.A. Marschall, and D.A. Swanson. 2004. The cost of dispersal: predation as a function of movement and site familiarity in ruffed grouse. *Behavioral Ecology* 15:469–476.

Index